THE PHYSICS OF THE EARTH'S CORE

AN INTRODUCTION

Related Pergamon Titles of Interest

Books

BATES
Geophysics in the Affairs of Man

CODATA BULLETIN
Geodesy

CONDIE
Plate Tectonics and Crustal Evolution, 2nd edition

GRIFFITHS & KING
Applied Geophysics for Geologists and Engineers, 2nd edition

HENDERSON
Inorganic Geochemistry

JONES
The Solar System

** MELCHIOR
The Tides of the Planet Earth, 2nd edition

Journals

Computers & Geosciences

Geochimica et Cosmochimica Acta

Nuclear Geophysics*

Planetary & Space Science

Vistas in Astronomy

Full details of all Pergamon publications/free specimen copy of any Pergamon journal available on request from your nearest Pergamon office

* New

** By the author, but not available under the terms of the Pergamon textbook inspection copy scheme.

"Do you realize you're eating the core!"

THE PHYSICS OF THE EARTH'S CORE

AN INTRODUCTION

by

PAUL MELCHIOR

Dr. Sc., Director of the Royal Observatory of Belgium, Bruxelles & Professor at the Catholic University, Louvain-la-Neuve

PERGAMON PRESS

OXFORD · NEW YORK · BEIJING · FRANKFURT
SÃO PAULO · SYDNEY · TOKYO · TORONTO

U.K.	Pergamon Press, Headington Hill Hall, Oxford OX3 0BW, England
U.S.A.	Pergamon Press, Maxwell House, Fairview Park, Elmsford, New York 10523, U.S.A.
PEOPLE'S REPUBLIC OF CHINA	Pergamon Press, Qianmen Hotel, Beijing, People's Republic of China
FEDERAL REPUBLIC OF GERMANY	Pergamon Press, Hammerweg 6, D-6242 Kronberg, Federal Republic of Germany
BRAZIL	Pergamon Editora, Rua Eça de Queiros, 346, CEP 04011, São Paulo, Brazil
AUSTRALIA	Pergamon Press Australia, P.O. Box 544, Potts Point, N.S.W. 2011, Australia
JAPAN	Pergamon Press, 8th Floor, Matsuoka Central Building, 1-7-1 Nishishinjuku, Shinjuku-ku, Tokyo 160, Japan
CANADA	Pergamon Press Canada, Suite 104, 150 Consumers Road, Willowdale, Ontario M2J 1P9, Canada

First edition 1986

Library of Congress Cataloging in Publication Data
Melchior, Paul J.
The physics of the earth's core.
Bibliography: p.
Includes index.
1. Earth—Core. I. Title.
QE509.M45 1986 551.1'12 85–32027

British Library Cataloguing in Publication Data
Melchior, Paul
The physics of the earth's core: an introduction
1. Geophysics
I. Title
551 QC802.A1

ISBN 0–08–032607–2 (Hardcover)
ISBN 0–08–032606–4 (Flexicover)

Printed in Great Britain by A. Wheaton & Co Ltd Exeter

Contents

Introduction

The motions inside the Earth's liquid core and their role in core–mantle coupling are of serious concern to astronomers and geodesists. These motions play a determinant role when defining and choosing a system of axes of reference with a view to precise positioning, space navigation, and other applications such as tectonic plates movements.

Such a system of reference is, by necessity, referred to the observatories situated upon different tectonic plates and is subject to oscillations with respect to space (precession-nutations) and to the global Earth's body itself (polar motions).

The liquid core plays a non-negligible role in these phenomena and measurements of irregularities in the Earth's rotation may offer new tools for advancing the study of core motions. Non-tidal decade fluctuations in the speed of rotation of the Earth are so important that only the Earth's core appears to be big enough to explain these phenomena by exchanges of angular momentum with the mantle, resulting from electromagnetic torques.

In order to investigate this difficult problem, astronomers and geodesists must be kept aware, through articles in the exponentially increasing number of high-level scientific publications, of the continuous progress being made in understanding the Earth's interior. This is not a simple task inasmuch as reading and understanding such papers presupposes a knowledge of special theories and a utilization of very peculiar parameters whose definitions are often not easy to trace in the usual treatises of classical physics.

This creates a barrier among the Earth science fields where agreement on a broad synthesis of our planet's properties and internal behaviour is necessary, if we are to build a mathematical, physical and chemical model of the Earth.

Seismology is a powerful technique that we can use in describing the radial distribution of rheological parameters such as density (ρ) and compressibility in the core, but it is much less successful in determining viscosity. It does not provide direct information about the movements inside the core even if it does allow us to construct a profile of the local Brunt-Väisälä frequency which characterizes the stratification. However, this parameter depends upon the factor $d\rho/dr$ which is not an observable quantity; this profile is, therefore, poorly controlled, uncertain and varies from one model of the Earth to another.

Gravimetry provides us with information about the topography of the internal boundaries but, as with seismology, the solution is not unique and does not say anything about the movements.

Geomagnetism only gives direct information about poloidal fields at the surface of the Earth and the downward extrapolation to the core boundary is uncertain for the higher-order terms. One can only speculate about the importance of toroidal fields with the constraint that such fields cannot produce very high heating by the Joule effect. Moreover "short-period" oscillations (period < 4 years) are unobservable because of the resistivity of the upper mantle.

There is, consequently, a large choice of possible dynamos while motions at the core surface can be described only very partially by observing, for example, the position of null flux curves at different epochs.

Seismology and geomagnetism indeed leave much freedom for speculation: "all modes of core convection remain speculative until some direct or indirect observational evidence is found" (Gubbins *et al.*, 1982). Any new information should therefore be most welcome.

Two new kinds of measurements, tightly bound together (nutations and tides) have revealed that the role of the liquid core is not negligible in these phenomena so that, if we could reach a still higher precision in these measurements, we could most probably derive new information about the core motions and core mantle boundary layers.

Tesseral diurnal tides and nutations are indeed two aspects of the same lunisolar attraction effects. Their amplitudes and frequencies are related by simple formulae (Melchior, 1971) and, in the last century, Sludsky demonstrated that a resonance could occur at a frequency very close to the diurnal frequency, while Poincaré (1910) developed an elegant formulation of this effect. There is resonance when the container oscillates at a period which is very close to one of the free oscillations of the fluid contained in it.

Jeffreys, Molodensky (1961) and very recently Wahr (1981) calculated such resonances for different Earth models, but only Wahr introduced completely the Earth's flattening and the Coriolis force in the formulations.

The resonance on Earth tides appears when one experimentally compares the major lunar diurnal waves (called Q_1, O_1) with the major solar diurnal waves (called P_1, K_1) and this leaves no doubt about the existence of this effect. But the strongest resonance appears on an extremely small solar diurnal wave (called ψ_1) which is associated with the annual nutation of the axes of inertia and rotation in space. If one could measure with sufficient precision the amplitude of this resonance on ψ_1 and, possibly, its phase, one could obtain exceptional information about the dissipation effects in the boundary layers at the core–mantle interface.

Observations in mines, made with small quartz pendulums, identified the liquid core resonance effects as early as 1960 (Melchior). These were later confirmed with spring gravimeters and extensometers. All of these instruments demonstrate, without any doubt, that the predicted resonance exists, but, unfortunately, they are not precise enough (see Chapter 5, Table 5.1) to determine this resonance to the four or five digits in amplitude which we require in order to identify the type of stratification and the viscosity of the core in the different models available, and in order to evaluate the damping.

New instruments are needed in this field. At present much hope has been placed in the superconducting gravimeters, recently put into operation, to detect internal gravity waves inside the liquid core. (Melchior and Ducarme, 1986).

In the meantime, direct observations of the nutations, and particularly of the annual nutation associated with the ψ_1 tidal wave, have been obtained from a new, powerful technique, Very long base interferometry (VLBI), which is extremely promising. A recent result, presented in a short summary published in *Eos* by Gwinn, Herring and Shapiro (1984), proposes not only a small correction to the resonance amplification of the annual nutation but, for the first time, gives a measure of the phase lag (about 0.8°).

Such effects, at this frequency, are due to some core oscillation with respect to the mantle, and this provides strong motivation to develop and improve all kinds of observations of these phenomena which are essentially related to the lunisolar attraction: the tides and their associated nutations. These new experimental results, due to the increased precision of modern instruments, are leading geodesists and astronomers to an even greater concern for the physics of the Earth's interior.

This book is the result of a request for an introductory course on the physics of the Earth's interior. I was asked to give such a course for geodesists in the summer of 1982, at Admont (Austria), by the University of Graz: the objective was above all to "open the barriers" by offering geodesists a condensed knowledge of those essential physical topics which would allow them to follow, with greater facility, the most recent developments in internal geophysics.

This book is an enlarged version of that course which was previously offered as a third cycle course at Louvain University and at the Royal Observatory of Belgium.

References will often be made to some recent results in order to better explain how physical theories can be used to describe the properties and behavior of the deep interior of the Earth. However, the reader should not consider these examples as firmly established conclusions even though they are probably very close to the truth.

The objective sought with this book is in no way to present an overview of the most recent results and progress in the different fields of internal geophysics, but to offer a kind of reference syllabus, or a key to a better understanding of the best theoretical papers in geophysics.

New results continuously flow in at an accelerated rate so that the interpretations and "conclusions" are still in a dynamic state.

Papers of major interest appear every month in several very high-standard scientific journals which are listed below, and which constitute a permanently updated bibliography on the subject:

Journal of Geophysical Research
(American Geophysical Union) vol. **54**, 1949 to vol. **80**, 1985.
(previously *Terrestrial Magnetism* 1896–1948; vols **1–53**)

Geophysical Journal of the Royal Astronomical Society
vol. **1**, 1958 to vol. **81**, 1985
(previously *Monthly Notices of the Royal Astronomical Society – Geophysical Supplement*)

Geophysical and Astrophysical Fluid Dynamics
vol. **9**, 1977 to vol. **32**, 1985
Gordon and Breach Science Publishers
(previously *Geophysical Fluid Dynamics*, vol. **1**, 1970 to vol. **8**, 1977)

Physics of the Earth and Planetary Interiors
Elsevier, Amsterdam
vol. **1**, 1967 to vol. **38**, 1985

Journal of Fluid Mechanics
Cambridge University Press, vol. **1**, 1956 to vol. **30**, 1985

These lessons aim to help young students read these papers where the basic formulae, definitions and theorems cannot, of course, be explained in detail by the authors in the limited space made available to them.

In addition to the journals listed above, a number of books exist which are given in the following bibliography.

Bibliography

Batchelor, G. K. (1967) *An Introduction to Fluid Mechanics.* Cambridge University Press.
Brush, S. G. (1979) Nineteenth-century debates about the inside of the earth solid, liquid or gas *Ann. Sci.* **36**, 225–254.
Brush, S. G. (1980) Discovery of the Earth's core. *Am. J. Phys.*, **48**, 705–724.
Brush, S. G. (1982) Chemical history of the Earth's core. *EOS*, **63**, 1185–1188.
Bullen, K. E. (1975) *The Earth's Density.* Chapman and Hall.
Chandrasekhar, S. (1961) *Hydrodynamic and Hydromagnetic Stability.* Clarendon Press.
Cole, G. H. A. (1984) *Physics of Planetary Interiors.* Adam Hilger Ltd.
Coulomb, J. and Jobert, G. (1976) *Traité de Géophysique Interne.* 2 vols. Masson.
Eckart, C. (1960) *Hydrodynamics of Oceans and Atmospheres.* Pergamon Press.
Greenspan, H. P. (1969) *The Theory of Rotating Fluids.* Cambridge University Press.
Jacobs, J. A. (1963) *The Earth's Core and Geomagnetism.* Pergamon Press.
Jacobs, J. A. (1975) *The Earth's Core.* Academic Press.
Jeffreys, H. (1970) *The Earth.* Cambridge University Press.

Landau, L. and Lifschitz, E. (1967) *Statistical Physics.* MIR, Moscow.
Lyttleton, R. A. (1953) *The Stability of Rotating Liquid Masses.* Cambridge University Press.
Melchior, P. (1982) *The Tides of the Planet Earth,* 2nd edn. Pergamon Press.
Moffat, H. K. (1978) *Magnetic Field Generation in Electrically Conducting Fluids.* Cambridge University Press.
Parker, E. N. (1979) *Cosmical Magnetic Fields: their origin and their activity.* Clarendon Press.
Pedlosky, J. (1979) *Geophysical Fluid Dynamics.* Springer-Verlag.
Roberts, P. H. (1967) *Introduction to Magnetohydrodynamics.* Longmans Green.
Roberts, P. H. and Soward, A. M. (1978) *Rotating Fluids in Geophysics.* Academic Press.
Schlichting, H. (1968) *Boundary-Layer Theory.* McGraw-Hill.
Shercliff, J. A. (1965) *A Textbook of Magnetohydrodynamics.* Pergamon Press.
Soward, A. M. (1983) Stellar and planetary magnetism *fluid mech. astrophys. geophys,* **2**, Gordon & Breach.
Squire, H. B. (1956) *Surveys in Mechanics* (ed. Batchelor and Davies). Cambridge University Press.
Stacey, F. D. (1977) *Physics of the Earth,* 2nd edn. Wiley.
Stacey, F. (1981) A thermodynamic approach to the problems of the Earth's interior. *IUGG Chronicle,* No. 149, 117–129.
Takeuchi, (1966) *Theory of the Earth's Interior.* Blaisdell.
Yih, C. (1965) *Dynamics of Nonhomogeneous Fluids.* Macmillan.
Zharkov, V. N. and Trubitsyn, V. P. (1978) *Physics of Planetary Interiors.* Tucson, Pachart.
NN. (1983) *Magnetic Field and the Processes in the Earth's Interior.* Czechoslovak Academy of Sciences, Prague.

References for the introduction

Gubbins, D., Thomson, C. J. and Whaler, K. A. (1982) Stable regions in the Earth's liquid core. Geophys. J. R. astr. Soc. 68, 241–251.
Gwinn, C. R., Herring, T. A. and Shapiro, L. I. (1984) Geodesy by Radio Interferometry: Corrections to the IAU 1980 nutation series. EOS, 65, 859.
Melchior, P. (1971) Precession-nutations and Tidal potential. Celestial Mech. 4, 190–212.
Melchior, P. and Ducarme, B. (1986) Detection of inertial gravity oscillations in the Earth's core with a superconducting gravimeter at Brussels. *Phys. Earth Plan. Int.* **42**, 129–34.
Molodensky, M. S. (1961) The theory of nutations and diurnal Earth tides. Comm. Obs. Roy. Belg. 188, S. Geoph. 58, 25.
Poincaré, H. (1910) Sur la Précession des corps déformables Bull. Astron. 27, 321–356.
Wahr, J. M. (1981) Body tides on an elliptical, rotating, elastic and oceanless Earth. Geophys. J. R. astr. Soc. 64, 705–728.

CHAPTER 1

The Model of the Earth

1.1 Shape and dimensions of the model

The model that we can propose and utilize for a description of a planet such as the Earth depends upon the precision which is requested when, using this model in a computer, the aim is to simulate future situations for predictions of natural phenomena.

For a long time such objectives were rather restricted: time determinations for economic and public use, predictions of eclipses, navigation of slow transport, which means that a few hundred metres were sufficient.

Recent progress in space geodesy and discoveries in geophysics (e.g. plate tectonics), associated with the availability of powerful computers and sophisticated measuring devices, have opened a new era with the hope—or the ambition—of making much more precise and significant simulations which could, perhaps—or probably—assist in earthquake prevision.

In this respect the requisite is centimetre-level precision for positioning terrestrial sites.

In such a situation the model to be constructed can no more be a simple reference surface with a simple equation like for an ellipsoid but a body approaching as close as possible the real Earth's body—that is an equipotential surface. This is not easy, for what concerns the surface of the Earth, it is much less easy for its interior.

A *standard Earth model* is a very large set of numbers, based upon few conditions or parameters:

(1) A scale length defined by the speed of light in vacuum:

$$c = 299\,792.458 \text{ km s}^{-1} \tag{1.1}$$

which defines the *metre* as the unit of length.

(2) A reference trihedron fixed at the centre of mass of the Earth with its axes Oz and Ox conventionally defined by the latitudes and longitudes attributed to a limited number of fundamental observatories situated upon different tectonic plates. These axes should be as close as possible to the principal axes of inertia of the Earth (see (6)) and are now defined in this way with a precision of about 0″2 (figure 1.1) but there are systematic discrepancies between the different systems of 0″5 or more.

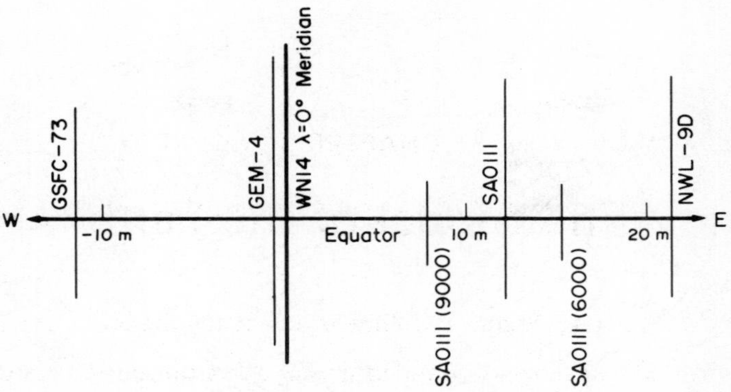

FIGURE 1.1 Dynamic zero meridians relative to the WN14 zero meridian.

(3) In this system of reference the Cartesian coordinates (x_i, y_i, z_i) of a large number (hundreds) of observatories situated all around the world, where all kinds of astrometric and satellite observations are or were made. This is the so-called *terrestrial polyhedron.*

(4) The geocentric gravitational constant of the Earth, GM where $G = 6.672\ \text{kg}^{-1}\ \text{m}^3\ \text{s}^{-1}$ and M is the Earth's mass. It is given, in principle, by the third law of Kepler:

$$G(M+m) = n^2 d^3 = 3\,986\,005\ 10^8\ \text{m}^3\ \text{s}^{-2} \tag{1.2}$$

where $m \sim 0$ when dealing with artificial satellites (n is the mean motion and d the semi-major axis of the orbit).

(5) The equatorial radius a of the best-fitting ellipsoid of revolution to the world mean sea level (mean Earth ellipsoid). Its value can be obtained by adjusting the heights of the summits of the terrestrial polyhedron above this ellipsoid to their heights directly measured by precise levellings starting from the reference maregraphs all around the world. It is now adopted to be:

$$a = 6\,378\,137\ \text{m} \tag{1.3}$$

(6) A list of purely numerical coefficients C_{nm}, S_{nm} representing the gravitational potential of the Earth into a development in Legendre polynomials $P_n^m(\cos\theta)$ usually written:

$$V = \frac{GM}{r}\left[1 + \sum_{n=2}^{\infty}\sum_{m=0}^{n}\left(\frac{a}{r}\right)^n (C_{nm}\cos m\lambda + S_{nm}\sin m\lambda)\,P_n^m(\cos\theta)\right] \tag{1.4}$$

here:

$$C_{10} = C_{11} = 0 \quad C_{12} = S_{12} = 0 \tag{1.5}$$

if condition (2) is fulfilled.

Despite lack of convergence the series (1.4) must evidently be truncated for practical applications. At present a list of 32 757 C_{nm}, S_{nm} coefficients is

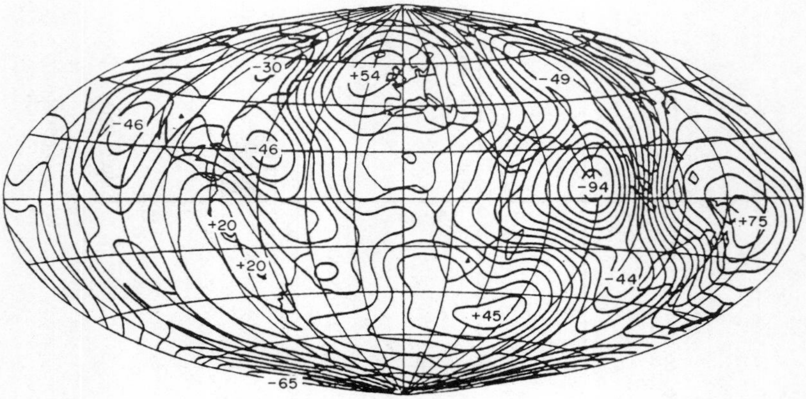

FIGURE 1.2 Geoid undulations with respect to a reference ellipsoid.

available on magnetic tape, which corresponds to about the order $n = 180$ (the number of coefficients to order n is $n^2 + 2n - 3$), and this may represent the geoid undulations with a precision of 2 or 3 metres or better (figure 1.2). The most important coefficient in this series is

$$C_{20} = -J_2 = -\frac{C-A}{Ma^2} \tag{1.6}$$

called the dynamical form factor of the earth. It is considered as one of the essential "constants" in geodesy. A huge number of observations of the LAGEOS satellite by laser ranging has allowed determination from the retrogradation of the orbit's nodes

$$J_2 = 10\,826.3 \times 10^{-7} \tag{1.7}$$

However it is systematically decreasing with time:

$$\dot{J}_2 = -3 \times 10^{-11}\ \text{year}^{-1}$$

an effect of Pleistocene deglaciation (Yoder *et al.*, 1983).
From this result follows also a numerical value of the geometrical flattening of the best-fitting ellipsoid:

$$f = \frac{a-c}{a} = 1/298.257\,222\,101 \tag{1.8}$$

through the Clairaut relation:

$$f = \frac{1}{2}q + \frac{3}{2}J_2 + \frac{9}{8}J_2^2 - \frac{3}{14}J_2 q - \frac{11}{56}q^2 \tag{1.9}$$

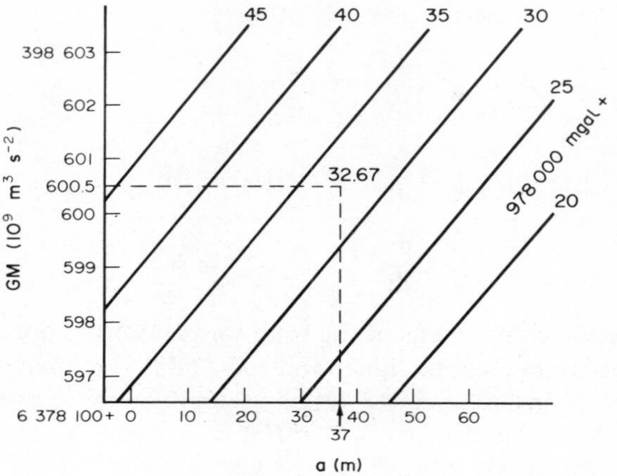

FIGURE 1.3 Relation between GM, a and the equatorial gravity.

where

$$q = \omega^2 a^3 / GM \tag{1.10}$$

and ω is the speed of rotation of the Earth, which is constant up to its seventh digit (see eqn. 1.13).

(7) A necessary check of the values obtained in this way consists in a comparison of a value of the acceleration of the gravity at a colatitude θ calculated with the Somigliana "international gravity formula":

$$\gamma(\theta) = \frac{GM}{a^2}\left[1 + f - \frac{3}{2}q + \left(\frac{5}{2}q - f\right)\cos^2\theta\right] \tag{1.11}$$

with absolute measurements of g made in the Bureau International des Poids et Mesures at Sèvres, in national metrological institutions like those at Leningrad, Teddington, Torino, Washington or along profiles like this one realized across Europe, from Tromsoe (Norway) to Catania (Italy) by the Istituto Metrologico Colonetti, Torino with a transportable absolute instrument.

The coherence of satellite results (1.2), (1.3), (1.11), with such ground measurements is demonstrated by figure 1.3. (Rigorous formulas may be found in "Geodetic reference system 1980", H. Moritz, *Bull. Geod.*, **58** (3), 388–398).

1.2 The rotation of the model of the Earth

The governing equations of the rotation of a solid body around its centre of mass are the well known *Euler equations* which, in a reference frame defined by the principal axes of inertia, are:

$$\begin{cases} A\dfrac{dp}{dt}+(C-A)\,qr = L \\ \\ A\dfrac{dq}{dt}-(C-A)\,rp = M \\ \\ C\dfrac{dr}{dt} = N \end{cases} \tag{1.12}$$

(L, M, N) are the components of the total torque exerted upon the earth by external forces, essentially the lunar and solar tidal potential; (p, q, r) are the projections of the instantaneous rotation vector $\bar{\omega}$ on the axes of inertia, with p and q extremely small $(p/r \sim q/r \sim 0''3)$ and $A \approx B$, C are the principal moments of inertia. Because $(B-A)$ is also extremely small, the product $(B-A)\,qp$ disappears from the third equation.

It can be easily demonstrated that, if the Earth's body has a cylindrical symmetry ($A = B$) and even if the deformations produced by the external potential W are purely elastic (Hooke body), $N = 0$ so that *if* C does not vary with time:

$$r = \text{constant} = \omega = 7.292\,115\;10^{-5}\;\text{rad s}^{-1} \tag{1.13}$$

Observations show that it is indeed constant up to its seventh digit, which is the mean velocity of rotation of the Earth and corresponds to a period of rotation of one sidereal day, that is 86 164.098904 seconds of universal time.

The relation (1.13) explains why, for centuries, the Earth's rotation could have been considered as the best available timekeeper (Laplace wrote that all the astronomy rests upon the hypothesis of the constancy of the Earth's rotation speed.*)

Variations of the speed of rotation

We know however that C is not constant, nor is N zero, so that this timekeeper can no longer satisfy the needs of our epoch, having at our disposal atomic clocks.

The variations of speed arise for a number of reasons, some of which are rather well controlled, others not:

(1) Periodic variations due to periodic variations of flattening caused by the zonal elastic earth tides. Their main periods are 13.7 and 27.5 days, 6 months, 1 and 18.6 years (see figure 1.4) with amplitudes of the order of milliseconds of time. These effects can be predicted with high precision.

* Laplace, *Mécanique Céleste, Tome V*

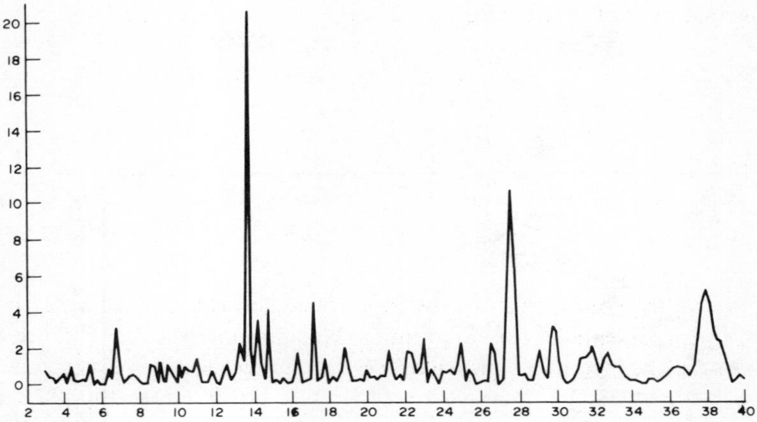

FIGURE 1.4 Spectrogram of (TU1–TUC) obtained by Djurovic from the BIH results of 1967–1974 (abscisse, the periods expressed in days).

(2) Periodic annual and semi-annual variations of speed due to exchange of momentum between the atmosphere and the "solid" Earth. These effects, which cannot be predicted with high precision, are now controlled practically in real time (see figure 1.5).

(3) A secular retardation which can be ascribed essentially to the friction of oceanic tides and, for a small additional amount, to the Earth tides. This effect is quite well controlled for the present millennia but, as its amplitude depends upon the continents' geographical pattern, the dimensions and depths of the oceans, it has been different during geological times. Its amplitude is, at present

$$\frac{d\omega}{dt} = -4.8\ 10^{-22}\ \text{rad s}^{-2} \quad \text{(figure 1.6)} \tag{1.14}$$

(4) Secular changes due to secular changes of C.

(5) "Sudden" accelerations or retardations (jerks) which are not predictable at all and seem to be correlated with changes in the geomagnetic field (Jacobs, 1975) (figure 1.7) and therefore could probably be monitored by torques exerted at the core–mantle boundary, deep inside the Earth: the causes of these intriguing phenomena are to be discovered in the deep interior of the Earth. *One aim of this book is to clarify such possibilities.*

For a body such as the Earth which, with its elastico-viscous mantle, its atmosphere, oceans and liquid core is essentially a deformable body, the *equations of Liouville*, which express the balance of momentum, are often more appropriate because the chosen axes of reference are not, and do not remain,

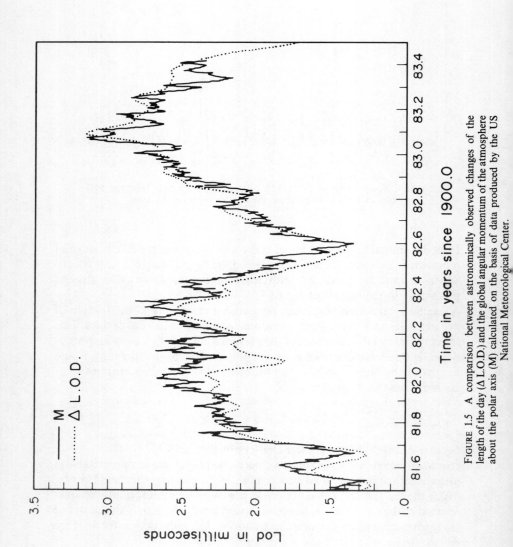

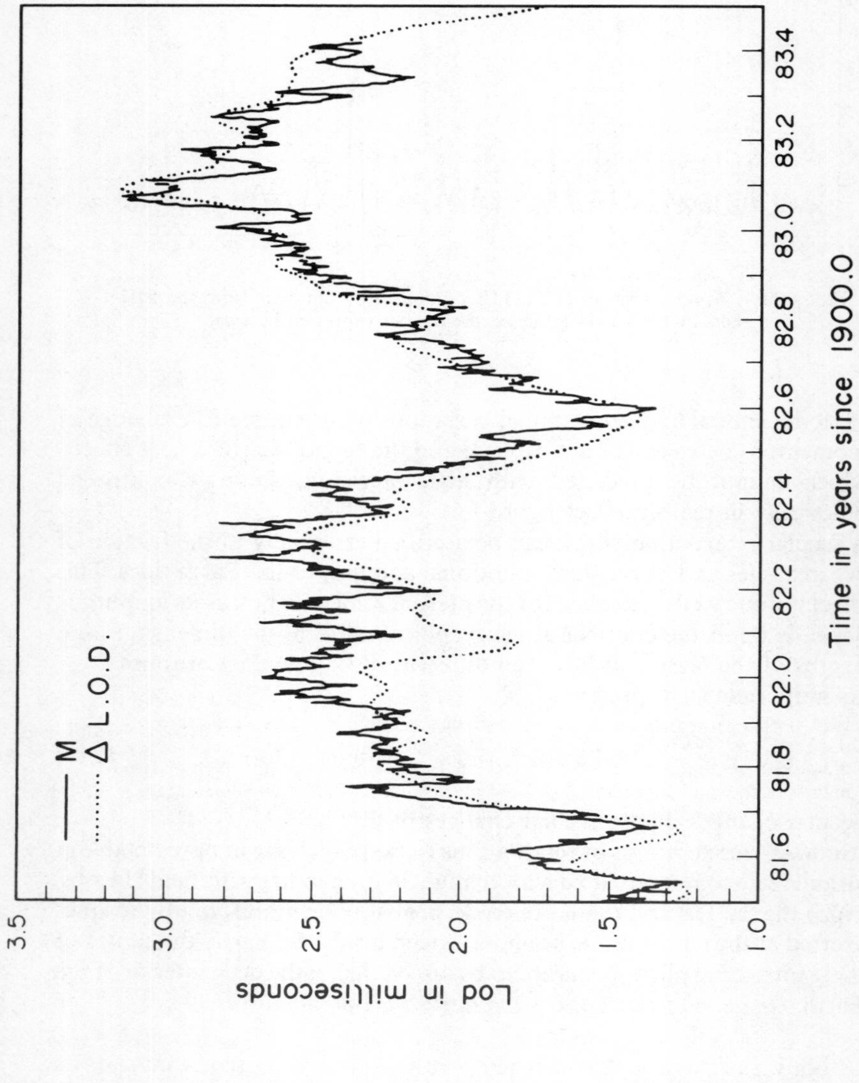

FIGURE 1.5 A comparison between astronomically observed changes of the length of the day (Δ L.O.D.) and the global angular momentum of the atmosphere about the polar axis (M) calculated on the basis of data produced by the US National Meteorological Center.

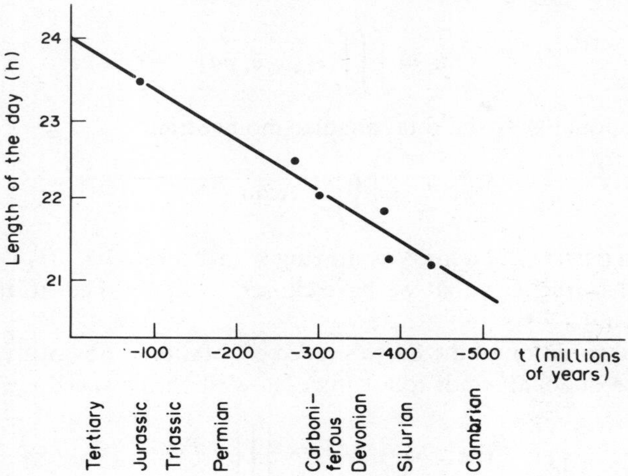

FIGURE 1.6.

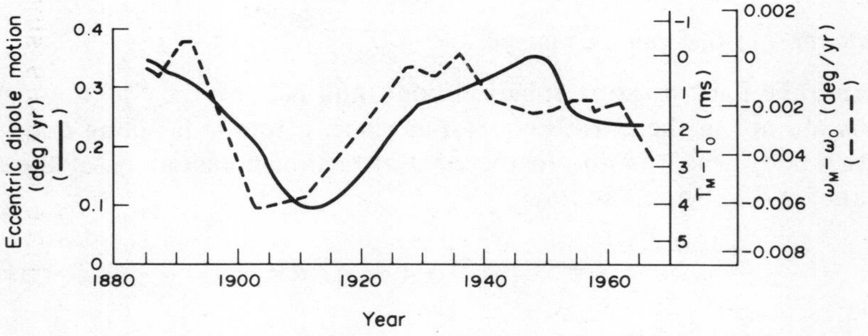

FIGURE 1.7 Comparison of astronomical and geomagnetic data on the speed of the Earth's rotation. The solid line represents the westward motion of the eccentric dipole relative to the mantle, the dashed line the eastward angular velocity of the mantle (with scales appropriate to deviations from both standard angular velocity and standard length of day). (After Jacobs, 1975).

principal axes of inertia during the time. The form of these equations is:

$$\frac{d}{dt}(C_{ij}\omega_j + h_i) + \varepsilon_{ijk}\omega_j(C_{kl}\omega_l + h_k) = L_i \tag{1.15}$$

where C_{ij} are the components of the inertia tensor

$\omega_j = p, q, r$ as before

$L_i = L, M, N$ as before

ε_{ijk} is the Levi–Civita antisymmetry tensor $(-1$ or $+1)$

while

$$h_i = \iiint_v \varepsilon_{ijk} x_j u_k \rho dV \tag{1.16}$$

are the components of the total angular momentum

$$\vec{h} = \iiint_v (\bar{r} \Lambda \bar{q}) \rho \, dV \tag{1.17}$$

of the Earth particles of density ρ, moving with a speed $\bar{q} = \bar{q}(u_k) = \bar{q}(u, v, w)$ with respect to the axes that we have chosen by fixing them to the tectonic plates [§1.1 (2)].

An annual variation of the Earth's speed of rotation is obviously created by the existence of zonal winds following east–west with a speed u_λ. This gives:

$$h_3 = \iiint_v (x_1 u_2 - x_2 u_1) \rho \, dV = \iiint_v \rho r^3 \cos^2 \phi u_\lambda dr \, d\phi \, d\lambda$$

which shows that, because of the $\cos^2 \phi$ coefficient, the effect of tropical winds is dominating.

Oscillations of the spin axis in space

Because the Earth is not a spherical body, and because the Moon and Sun orbits are not in the Earth's equatorial plane, a torque is acting upon the Earth's body which is due to the tidal attraction potential W of the two external bodies, Moon and Sun:

$$\vec{N} = - \iiint_v (\bar{r} \Lambda \, \overline{\text{grad}} \, W) \rho d V \tag{1.18}$$

which can be written*:

$$\vec{N} = \oint\!\!\oint_S (\vec{n} \Lambda \vec{R}) \rho W dS + \iiint_v (\bar{r} \Lambda \, \overline{\text{grad}} \, \rho) W dV \tag{1.19}$$

two terms corresponding respectively to the geometrical and to the dynamical flattening of the Earth's body.

We have demonstrated elsewhere (Melchior, 1971, 1980, 1983) that only the tesseral diurnal tides contained into the tidal potential W produce non-zero torques $\vec{N}$.

The result is a precession and an infinite number of nutations of the principal axis of inertia in space but, as the axis of rotation always remains extremely close to the axis of inertia (their angular separation in space is about

* This results from the relation

$$\text{rot}\,(\rho W \bar{r}) = \rho W \, \text{rot}\, \bar{r} - (\bar{r} \Lambda \, \overline{\text{grad}} \, \rho W) = -(\bar{r} \Lambda \, \overline{\text{grad}} \, W) \rho - (\bar{r} \Lambda \, \overline{\text{grad}} \, \rho) W$$

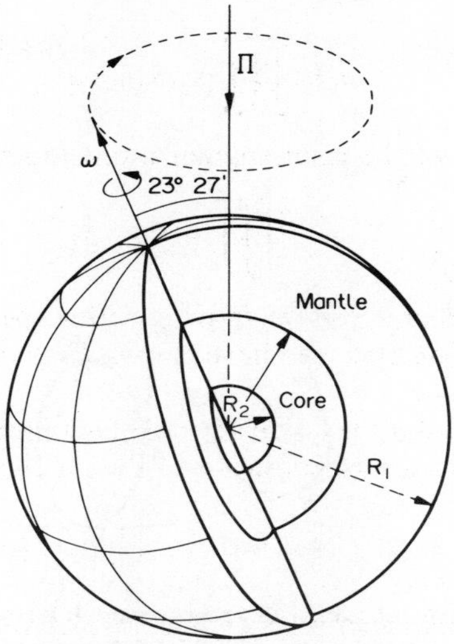

FIGURE 1.8 Lunisolar precession: the instantaneous axis of rotation $\vec{\omega}$ describes in space, counterclockwise, a circular cone of 23°27′ amplitude around the pole π of the ecliptic. The period is 25 800 years.

0″001), the astronomers observe a very large size precession (angular amplitude 23°27′, period 25 800 years) and a number of main nutations of the *axis of rotation* in space.

An obvious question, often raised in the past century, is to know how the liquid core reacts to such oscillations of its container, the mantle, and whether it can experience resonance effects upon some frequencies of oscillation.

In many textbooks of astronomy or geodesy, drawings are presented depicting the combined effects of precession and nutation as regular sinusoidal oscillations. This is misleading because the *annual* precession is 50″26 while the principal nutation is only 9″2 in 18.6 years, the semi-annual nutation is 0″5 in 6 months, the fortnightly nutation is 0″098 in 14 days so that the real nutation effect looks epicycloidal as on figures 1.8 to 1.11 with cusps and retrogressions. This creates non-linearities in the liquid core reactions. Unfortunately such "short" periods cannot be observed from the geomagnetic field because of the resistivity of the mantle (see chapter 4).

Astronomical observations performed over several centuries have allowed determination of the coefficient in the torque resultant which is called the precession constant:

$$H = \frac{C - A}{C} = 1/305.43738 \qquad (1.20)$$

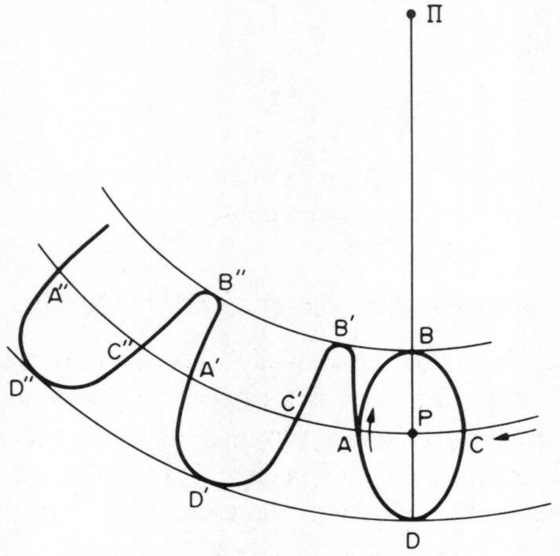

FIGURE 1.9 Principal nutation of 18.6 years period. It is counterclockwise as the lunar node movement is retrograde. During a precession period of 25 800 years there are 1380 nutation oscillations DD' is about 940" while PB is 9"20, PA is 6"85. It is therefore difficult to represent this oscillation at correct scale.

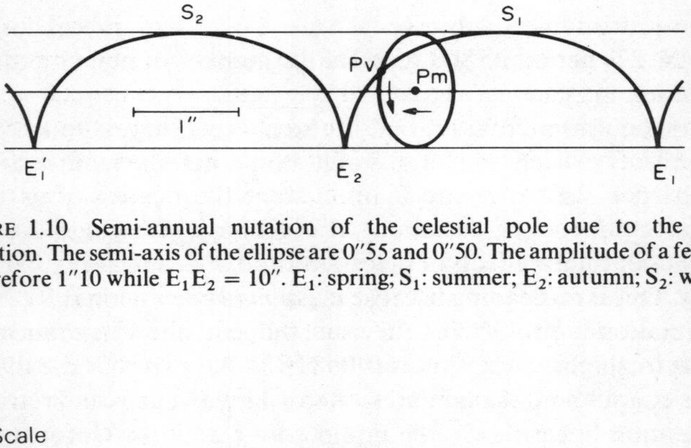

FIGURE 1.10 Semi-annual nutation of the celestial pole due to the solar attraction. The semi-axis of the ellipse are 0"55 and 0"50. The amplitude of a festoon is therefore 1"10 while $E_1E_2 = 10''$. E_1: spring; S_1: summer; E_2: autumn; S_2: winter.

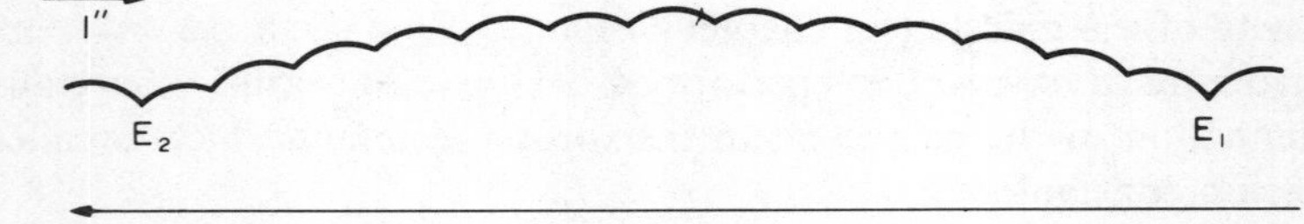

FIGURE 1.11 Fortnightly nutation superposed on the semi-annual nutation (13 festoons within 6 months). Semi-axis of the ellipse are 0"088 and 0"081.

From (1.7) and (1.20) one then obtains:

$$\frac{C}{Ma^2} = 0.330676 \tag{1.21}$$

a parameter which is an essential constraint for any model of density distribution inside the Earth.

Oscillations of the spin axis with respect to the Earth's body

(1) It is well known from the beautiful Poinsot geometrical representation (1854) that the instantaneous axis of rotation of a solid body, rotating around its centre of mass, cannot remain rigorously fixed inside the body if it oscillates in space, and vice-versa. The "mouvement à la Poinsot" consists in cones rolling on each other without slipping, the contact line being the rotation axis which describes one cone in space and the other inside the body.

A consequence is that, in correspondence with the precession and the nutations, one must consider the existence of *forced* oscillations of the axis with respect to the Earth's body with nearly diurnal periods and amplitudes of 0″009, 0″006, 0″003 and so on; quite small perturbations but which may also be subject to resonance in the liquid core.

(2) There exists a more important phenomenon which is a free oscillation excited by causes internal to the system: atmospheric and oceanic impulses, possibly earthquakes.

The equations of this "free motion" are simply the Euler equations (1.12) without second member. The third one being obviously decoupled from the two others, these two are easily integrated, giving a circular movement with frequency $\omega(C - A)/A$ which corresponds to a period of 305 days, the "Euler period". This is unrealistic because the Earth is deformable and considerations about its elasticity made by Newcomb (1892) led him to explain the celebrated discovery by S. C. Chandler (1891) of a 427-day period. The respective roles of the elastic mantle, the liquid core and the oceans make the problem of the exact value of the Chandler period very difficult to calculate, while the results of 85 years observations by a selected group of observatories (International Latitude Service, thereafter International Polar Motion Service) do not clarify the question because of the small amplitude—about 0″3 (see figure 1.12)—of this motion and of the presence of perturbations of different kinds in the measurements. It is hoped that the new techniques, essentially Very long base interferometry, will help to solve the question when a sufficiently long series of data will have been obtained.

The other serious still unsolved question is that of the damping of this Chandler free motion and its source of regeneration: as said before it has been observed continuously since 1900 and does not disappear.

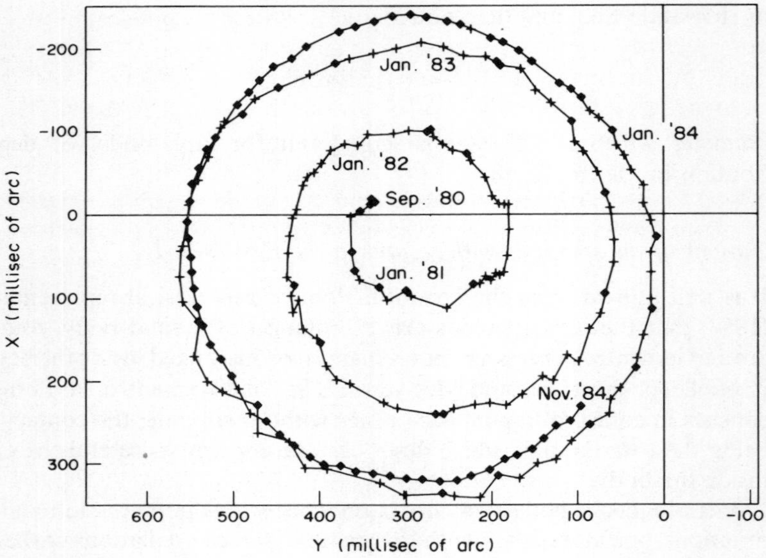

FIGURE 1.12 VLBI determinations of the pole (Carter *et al.*, 1985)

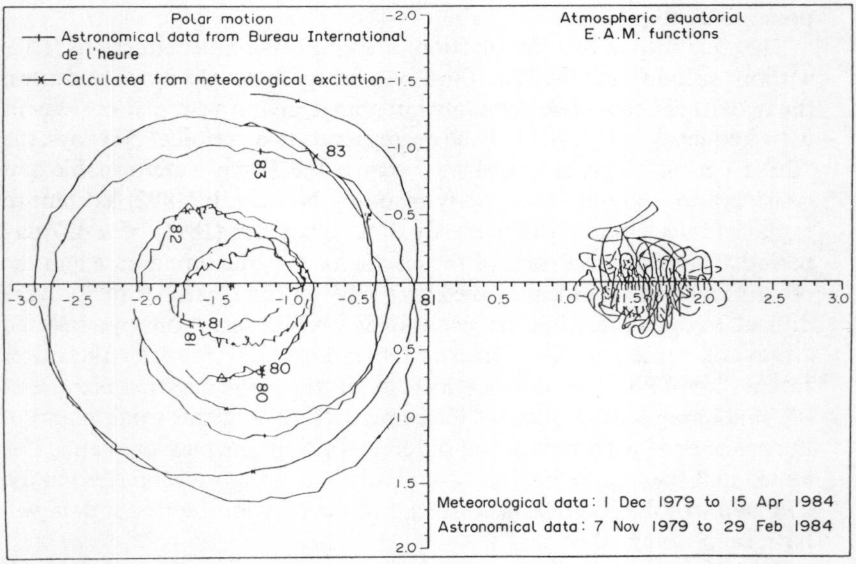

FIGURE 1.13 From Geophysical Fluid Dynamics Laboratory. Meteorological Office, Bracknell, United Kingdom

Meteorological excitation seems to be responsible, as shown on figure 1.13 by Barnes *et al.* (1983).

(3) A mode other than the "Chandler wobble" is possible because of the obvious fluidity of the Earth's core (see §1.3). It essentially consists in a nearly diurnal free wobble, the period of which depends upon the geometrical flattening of the core and is very close to the sidereal period of rotation. The existence of such a mode explains the possibility of resonance for those forced nutations and associated tesseral tides having frequencies close to the frequency of the mode.

1.3 Gravity and pressure inside the Earth

Gravity inside the Earth in the spherical approximation

Let us introduce the mean density of a sphere of radius r:

$$D_r = \frac{3}{r^3}\int_0^r \rho(x)x^2\,dx \tag{1.22}$$

and calculate its variation in function of r:

$$\frac{dD_r}{dr} = -\frac{9}{r^4}\int_0^r \rho(x)x^2\,dx + \frac{3}{r^3}\rho_r r^2 = -\frac{3}{r}(D_r - \rho_r). \tag{1.23}$$

Now

$$g_r = \frac{4\pi G}{3} D_r r \tag{1.24}$$

so that

$$\frac{dg_r}{dr} = \frac{4\pi G}{3}\left(D_r + r\frac{dD_r}{dr}\right) = 4\pi G\left(\rho_r - \frac{2D_r}{3}\right) \tag{1.25}$$

which is the expression of the *theorem of Saigey* stating that, as at the surface of the Earth $(r = a)$ $D_a = 5.52$, one has:

$$\frac{dg_a}{dr} = 4\pi G(\rho_a - 3.68) < 0 \tag{1.26}$$

g begins to increase when one penetrates inside the mantle.

At any point situated upon an internal surface of radius $r = b$ we have

$$g_b = \frac{4\pi G}{b^2}\int_0^b \rho r^2\,dr \tag{1.27}$$

(a reminder that an ellipsoidal envelope exerts no attraction on a point inside (Lagrange)). This easily explains the peculiar behaviour of the $g(r)$ function inside the Earth as shown on figure 1.14. Typical values are shown in Table 1.1 for depths where major density discontinuities appear.

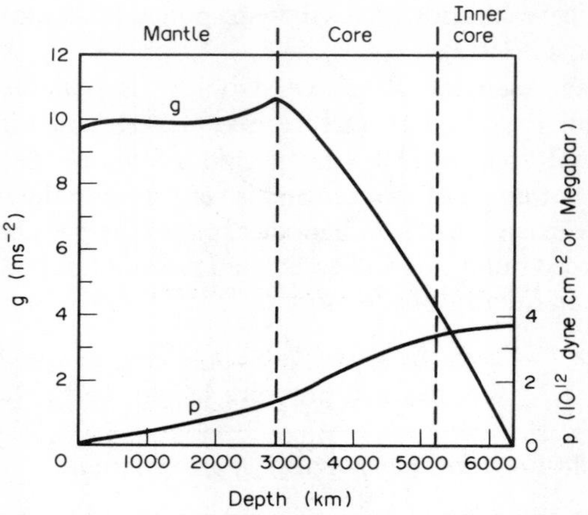

FIGURE 1.14 Variations of gravity (g) and pressure (p) for a model of Bullen (1967).

TABLE 1.1

Depth $(a-r)$	g (in gals)	
100 km	989	
400 km	998	
700 km	1000	
2898 km	1037	(core–mantle boundary)
5121 km	590	(inner core–core boundary)

(1 gal = 1 cm s^{-2})

Hydrostatic pressure inside the Earth

The law of the hydrostatic equilibrium:

$$\overline{\text{grad}}\, p = \rho\, \overline{\text{grad}}\, V \tag{1.28}$$

which, in the spherical approximation, can be written:

$$\frac{dp\,(r)}{dr} = -\rho\,(r) g(r) \tag{1.29}$$

can easily be integrated by using eqn. (1.24).

$$p(r) = 4\pi G \int_r^a \rho\,(r) \frac{dr}{r^2} \int_0^r \rho\,(x) x^2\, dx \tag{1.30}$$

(see figure 1.15) which gives the $p\,(r)$ profile across the Earth's interior. This is shown on figure 1.14.

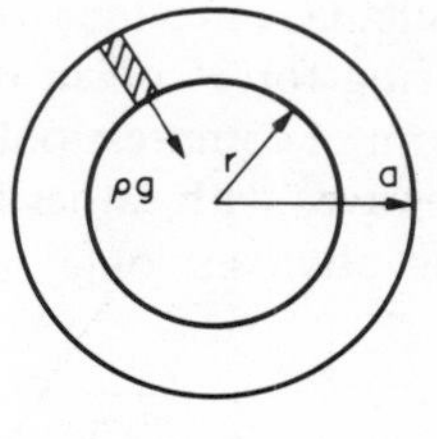

FIGURE 1.15 Hydrostatic pressure at radius *r*.

With the most recent models (see Table 1.4) one obtains for the pressure:

TABLE 1.2

Depth (km)	Radius (km)	Pressure (megabar)	
420	5951	0.141	crystallographic discontinuity
670	5701	0.239	crystallographic discontinuity
2886	3485	1.354	core–mantle interface
5154	1217	3.289	inner–outer core interface
6371	0	3.632	centre of mass

1 megabar = 10^{12} dynes cm^{-2} = 10^{11} pascal.

It had been suggested by Elsasser and Isenberg (1949) that, as in Fe atom the level 3d is not completely filled, a transition from $3d^6 4s^2$ to $3d^8$ could be possible under the effect of high pressure.

Bukowinski showed in 1976 that the pressure in the core is *not* sufficient to produce such an electronic collapse of iron ($3d^6 4s^2 \rightarrow 3d^8 4s$)*. If it happened, it would have as a result an increase of the electrical resistivity because 3d electrons have lower mobility than 4s electrons and consequently could create difficulties for the geodynamo equations (see §4.12).

1.4 Classical seismological model for the internal structure of the Earth

The rheological behaviour of solids and liquids, submitted to an oscillatory mechanical forcing, results from the resistances that matter offers to deformations.

* Iron family structure:

		K	L		M			N			
		1s	2s	2p	3s	3p	3d	4s	4p	4d	4f
26	Fe	2	2	6	2	6	6	2	–	–	–
27	Co	2	2	6	2	6	7	2	–	–	–
28	Ni	2	2	6	2	6	8	2	–	–	–

(s: sharp; p: principal; d: diffuse; f: fundamental).

Incompressibility resists a change in volume, rigidity a change of shape. The resulting forces act as restoring forces of an oscillating system capable of transporting energy in the form of compressional, longitudinal waves (sound waves) and shear transverse waves with rather high speeds.

In the Earth's mantle one observes longitudinal waves (compression *P* waves) of speed

$$V_P = \sqrt{\frac{\lambda + 2\mu}{\rho}} = \sqrt{\frac{k + \frac{4}{3}\mu}{\rho}} \sim 8 \text{ to } 13.7 \text{ km s}^{-1} \qquad (1.31)$$

and transverse waves (*S* shear waves) of speed

$$V_S = \sqrt{\frac{\mu}{\rho}} \sim 5 \text{ to } 7.5 \text{ km s}^{-1} \qquad (1.32)$$

λ and μ are the Lamé parameters, k the incompressibility or bulk modulus, μ the rigidity modulus and ρ the density, but this is valid only for isotropic materials.

A liquid opposes no resistance to a change of shape and this does not provide any restoring force; therefore, it cannot sustain transverse waves: such waves are not observed in the Earth's core.

One of the essential aspects of theoretical and experimental seismology is the detailed study of the propagation of elastic waves inside the Earth.

By treating the Earth as an idealized medium, seismologists soon discovered the main features of the Earth's interior which are represented on figure 1.16. Reflection and refraction of the longitudinal and transverse seismic waves allow one to determine with precision the positions of discontinuities in velocity while the observed speed of these waves gives informations about the density ρ, incompressibility k and rigidity μ in function of the radius r (figure 1.17). Seismologists are now developing the theory of propagation of the seismic waves in realistic media, i.e. media with lateral heterogeneities.

PnKP waves in the core

Most spectacular are the observations of multiple reflexions of the P waves inside the liquid core, called P*n*KP (*n* for the number of reflexions, K for kern = core).

These longitudinal waves are the result of a number of *n* successive reflexions inside the core boundary, as illustrated by figure 1.18. Up to 8 reflexions have been observed by *seismic arrays* which are ensembles of up to 500 seismographs distributed along prescribed patterns (Adams, 1972; Buchbinder, 1972; Qamar and Eisenberg, 1974, Wright, 1973). The radius of the core (3485 ± 3 km) is such that, as shown by figure 1.19, stations at distance 60° from the epicentre can observe P4KP and P7KP with a time delay of

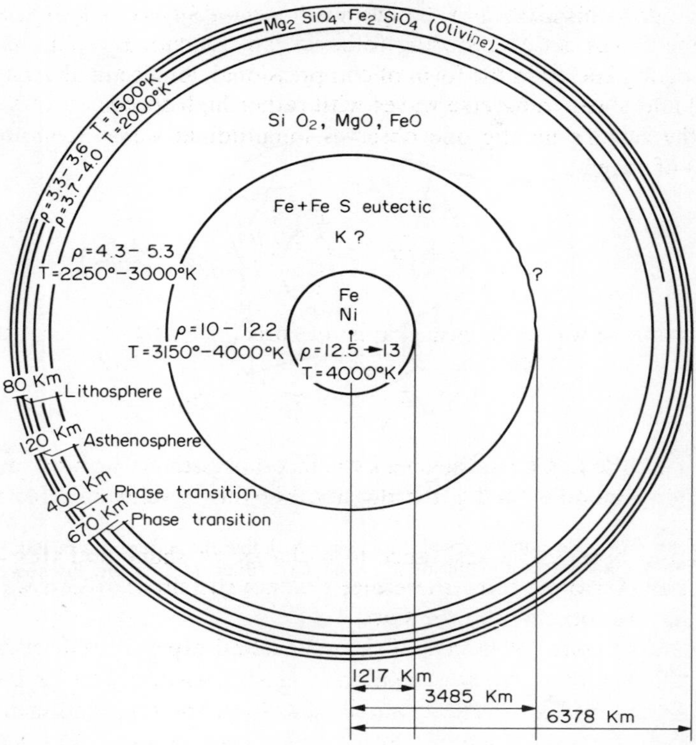

FIGURE 1.16 Classical representation of the structure of the Earth's interior. (A more realistic model involves lateral hetcrogeneities, i.e. non-radial distributions of density. The recent development of tomography reveals very important heterogeneities in the mantle.)

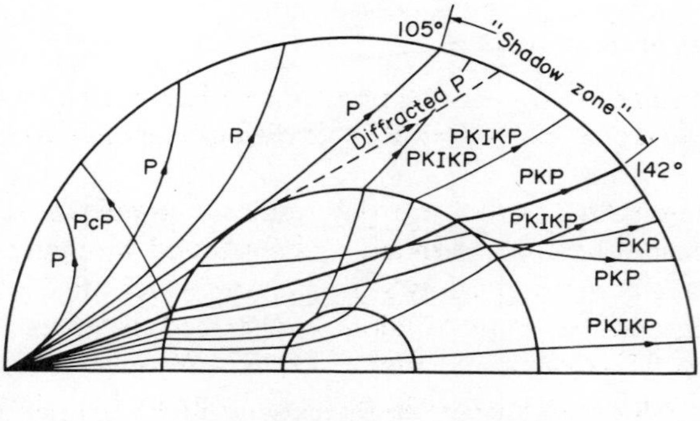

FIGURE 1.17 Different paths for longitudinal waves.

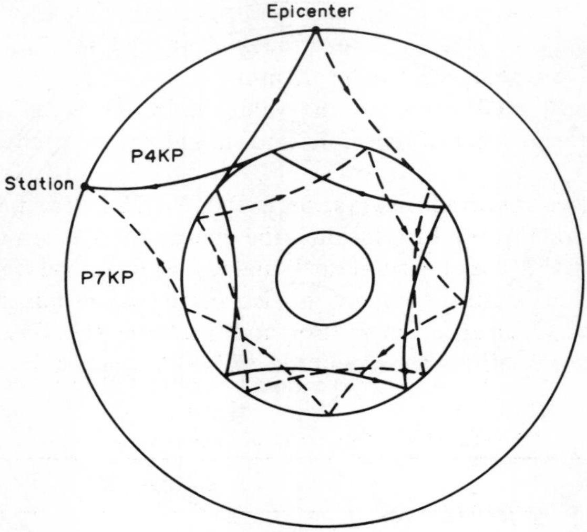

FIGURE 1.18 The dimensions of the core are such that the emergences of P*n*KP for $n = 4, n = 7$ practically coincide at $\Delta = 60°$ (according to Qamar and Eisenberg, 1974).

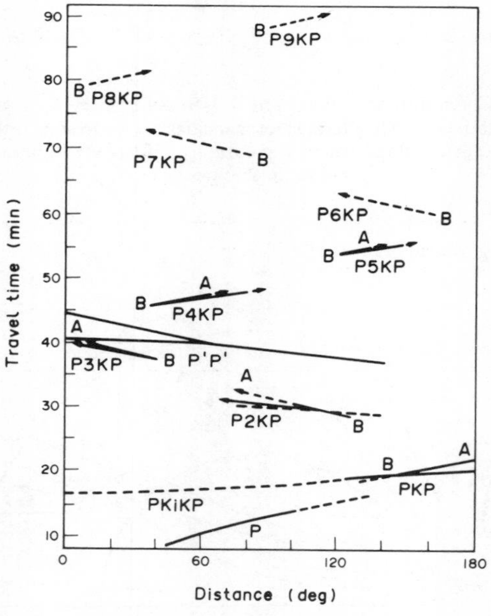

FIGURE 1.19 Times of arrival and angular epicentral distances of P1KP to P9KP (according to Qamar and Eisenberg, 1974).

respectively 47 min and 70 min, stations at distance 100° can observe P2KP after 30 min and P9KP after 90 min; stations at distance 130° can observe P5KP after 55 min and P6KP after 62 min.

Obviously any small error on the value of the core radius should be multiplied by the number of reflexions and should appear in the propagation times.

The most often reported observation is P4KP/P7KP (frequency ~ 1 Hz) from nuclear underground explosions (the energy release is more simple in that case than in the case of natural earthquakes). Their arrival is abrupt, which reconfirms that the mantle–core interface is a sharp discontinuity: its thickness is surely less than 2 km according to Bolt and Qamar (1972). The amplitude of P7KP is about one-third of the amplitude of P4KP, which shows that the core

TABLE 1.3 *Mean radius of the Earth's core*

Authors		Method	r (km)	
Jeffreys		PcP, ScS	3473	±2.5
Kogan	1960	PcP	3486	
Taggart Engdahl	1968	PcP-P nuclear explosions	3477	±2.0
Bolt	1970	PcP-P	3479	±2
Hales Roberts	1970	ScS-S	3489.92	±4.66
			3486.10	±4.59
Buchbinder	1971	PKP, P2KP, P*n*KP nuclear explosions	3479	±2
Wang	1972		3481	
Gilbert, Dziewonski, Brune	1973	PcP, ScS excluded	3482–3485	
Dziewonski, Gilbert	1973	DG579	3483.6	
Jordan, Anderson	1974	PcP-P	3485	
Engdahl, Johnson	1974	PcP	3485	±3
Gilbert	1974		3484.9	
Dziewonski, Haddon	1974		3485	±3
Dziewonski, Hales, Lapwood	1975	free oscillations and earthquakes	3485.7	

PcP, ScS are reflections outside the core boundary.

TABLE 1.4 *Mean radius of the inner core*

Authors		Method	r (km)	
Jeffreys	1939	PKP	1250	
Bolt	1964	PKiKP	1216	
Bolt, O'Neil, Qamar	1968	PKiKP	1220	
Buchbinder	1971		1226	±10
Bolt	1972		1216	±5
Qamar	1973		1213	
Dziewonski, Gilbert	1973	DG579	1228.9	
Engdahl, Flinn, Massé	1974	PKiKP-PcP	1227.4	±0.6
Dziewonski, Hales, Lapwood	1975	free oscillations and earthquakes	1217.1	

quality factor is high. (This agrees with the very low viscosity determined by Gans (1972): 10^{-3} stokes). A high number of reflections allows evidently determination of the mean core radius with high precision. The radii of the outer and inner cores are presently known with a precision of 1 or 2 km, as shown in Tables 1.3 and 1.4.

The Adams Williamson equation

Early seismologists tried to construct a density model for the Earth's interior on the basis of the sole velocities of longitudinal P and transverse S waves, as given by equations 1.31 and 1.32. From these velocities we derive the *seismological parameter* which is an observed quantity:

$$\phi = V_P^2 - \frac{4}{3} V_S^2 = \frac{k}{\rho} \ (\mathrm{km}^2\,\mathrm{s}^{-2}) \tag{1.33}$$

Let us consider a *Hooke body* in hydrostatic equilibrium. We may write the internal pressure as

$$p = -X_x = -Y_y = -Z_z = -(\lambda\theta + 2\mu\, e_{xx}) \tag{1.34}$$

where

$$\theta = \frac{\partial u}{\partial x} + \frac{\partial v}{\partial y} + \frac{\partial w}{\partial z} = 3e_{xx} \tag{1.35}$$

is the cubical dilatation and λ, μ the Lamé constants (u, v, w being the components of the displacement). Thus:

$$-p = (3\lambda + 2\mu)\, e_{xx}, \quad \theta = -\frac{p}{\lambda + \frac{2}{3}\mu} \tag{1.36}$$

and the adiabatic modulus of incompressibility may be defined by

$$k_S = -\frac{p}{\theta} = \lambda + \frac{2}{3}\mu \tag{1.37}$$

Let us now consider the adiabatic transformation*

$$-\frac{1}{k_S}\Delta p = \frac{\Delta v}{v} = -\frac{\Delta\rho}{\rho}$$

$$k_S = \rho \frac{dp}{d\rho} \tag{1.38}$$

and, to eliminate the pressure, use again the hydrostatic hypothesis (1.29).

$$k_S/\rho = dp/d\rho = (dp/dr)\cdot(dr/d\rho)$$

* From eqn (2.5) where dS is put equal to zero (see also eqn. 2.25).

giving:

$$(d\rho/dr) = -g \frac{\rho}{(k_S/\rho)} \tag{1.39}$$

which can be written:

$$\boxed{\frac{d\rho}{dr} + \rho \frac{g}{\phi(r)} = 0} \tag{1.40}$$

this is the Adams Williamson equation used by Bullen in 1936 to construct his first models of the Earth's interior (Melchior, vol. 3, 1972).

Free oscillations

The observation of free oscillations of the whole Earth's body on the occasion of very strong earthquakes (magnitude > 7.5), which succeeded for the first time with the Chile earthquake in 1958, has offered the seismologists an exceptional tool for constructing new refined models of the Earth's interior. Two types of oscillations are observed (see Appendix p. 243): spheroidal modes (with a radial component, observable with long period instruments like tidal gravimeters) and torsional modes which have no radial components and are observed only in the horizontal directions (with long period clinometers or extensometers).

The theory of free oscillations has been developed according to the same lines as the theory of Earth tides (which concerns only spheroidal components in the case of a spherical non-rotating elastic isotropic model: the SNREI model Earth) by Alterman, Jarosch and Pekeris (1959) and by Molodenskii (1961) independently.

Quite recently J. Wahr (1981) has developed a more general theory for an elliptic rotating elastic isotropic earth which, as a basic result, has a coupling of an infinite series of toroidal and spheroidal modes. For practical applications this necessitates a truncation. This should not create serious problems because the successive terms follow a hierarchy which is proportional to increasing powers of a very small coefficient (the flattening).

However in certain cases, for low-frequency modes, the truncated expansion appears to give rise to spurious features (Friedlander, 1985).

Models

The models of the Earth's interior are built on the basis of seismic observations and give three parameters in function of the radius r: the density $\rho(r)$, the velocity of compression waves $V_P(r)$ and the velocity of shear waves $V_S(r)$ from which the moduli of incompressibility $k(r)$ and rigidity $\mu(r)$ can be deduced as well as the Poisson coefficient $\sigma(r)$ and the Young modulus $E(r)$:

$$\left.\begin{aligned} \lambda &= \rho(V_P^2 - 2V_S^2) \\ \mu &= \rho V_S^2 \\ k &= \rho(V_P^2 - \tfrac{4}{3}V_S^2) \\ \sigma &= \tfrac{1}{2}\lambda/(\lambda+\mu) = \tfrac{1}{2}(V_P^2 - 2V_S^2)/(V_P^2 - V_S^2) \\ E &= \mu(3\lambda + 2\mu)/(\lambda+\mu) = \rho V_S^2(3V_P^2 - 4V_S^2)/(V_P^2 - V_S^2) \end{aligned}\right\} \quad (1.41)$$

these parameters are given at discrete intervals (every 100 km for example) or under the form of piecewise continuous analytical functions of the radius (polynomials) which permits easy calculation of their gradients.

This last presentation was adopted by Dziewonski, Hales and Lapwood (1975) for their three PEM models.

These three PEM models are based upon observations of free oscillations for 1064 normal modes and 246 travel times of body waves. They are identical below the depth of 420 km but, above, they reflect the different properties of the oceanic upper mantle (PEM-O), of the continental upper mantle (PEM-C) and of an average structure (PEM-A).

For depths greater than 670 km the density distribution is consistent with the Adams Williamson equation to within 0.2% maximum deviation.

The density jumps are given in table 1.6 for the most recent models.

The small jumps at 420 and 670 km are due to a change of crystallographic structure, under the effect of pressure of the orthosilicate mixtures which constitute the Earth's mantle. An important feature to point out here is that one has never observed any earthquake having a focus deeper than 670 km (see figure 1.22). Moreover there is a sharp burst of seismic activity just above 670 km.

The large jump at 2886 km depth is obviously due to a chemical difference: the abrupt transition from silicates to liquid iron.

TABLE 1.5 *The PEM-A model*

Radius range (km)	Depth range (km)	Region
0–1217	6371–5154	inner core
1217–3485	5154–2886	outer core
3485–5701	2886–670	lower mantle
5701–5951	670–420	transition zone
5951–6151	420–220	below LVZ
6151–6291	220–80	low velocity zone (LVZ)
6291–6352	80–19	above LVZ
6352–6357	19–14	lower crust
6357–6368	14–3	upper crust
6368–6371	3–0	ocean

The oceanic structure has a 160 km thickness LVZ while the continental structure has only a 100 km thick LVZ.

TABLE 1.6 *Density and density jumps at the major discontinuities inside the Earth.*

Depth (km)	Radius (km)	Models: B_1		*1066 A*		*1066 B*		*PEM*		*PREM*	
420	5951	3.58 3.80	0.22	3.712	0	3.605 3.746	0.141	3.553 3.768	0.215	3.543 3.724	0.181
670	5701	4.05 4.38	0.33	4.208	0	4.100 4.372	0.272	4.077 4.377	0.300	3.992 4.381	0.389
2886	3485	5.58 9.90	4.32	5.528 9.914	4.386	5.563 9.977	4.414	5.550 9.909	4.359	5.566 9.903	4.337
5154	1217	12.11 12.28	0.17	12.153 13.021	0.868	12.057 12.620	0.563	12.139 12.704	0.565	12.166 12.764	0.598
Centre 6371	0	12.58		13.421		13.077		13.012		13.088	

Note: The depths of discontinuities differ slightly (about 1 km but more for the inner core) from one model to the other.

References:	B_1	Jordan and Anderson, 1974
	1066 A, 1066 B	Gilbert and Dziewonski, 1975
	PEM	Dziewonski, Hales and Lapwood, 1975
	PREM	Dziewonski and Anderson, 1981
Other models:		
	B_2	Bullen, 1965
		Bullen A, Bullen B, 1967
	HB	Bullen and Haddon, 1973
	DG 579	Dziewonski and Gilbert, 1973
	C2	Anderson and Hart, 1976

This strong change of density characterizes the core–mantle interface, it looks very abrupt so that this boundary acts as a quite perfect reflector for the seismic waves. It has not been possible to evaluate any thickness for this interface. There is, however, at the bottom of the lower mantle, a layer called D″ by Bullen, where an inversion in the speed of P waves appears. We will later refer to this sheet to show that it can be interpreted as a thermal boundary layer (see figure 5.2).

As we shall see later, the jump at 5154 km, which is about 0.6 $g\,cm^{-3}$, is a critical parameter for the energetic budget of the geodynamo (chapter 4).

The recent models are based upon a high number of normal mode periods of free oscillations and travel time observations. As an example the PREM model involves 1000 normal mode periods and 12 years of the International Seismological Center phase data, that is $1.75 \cdot 10^6$ travel time observations for P and S waves (figures 1.20 and 1.21).

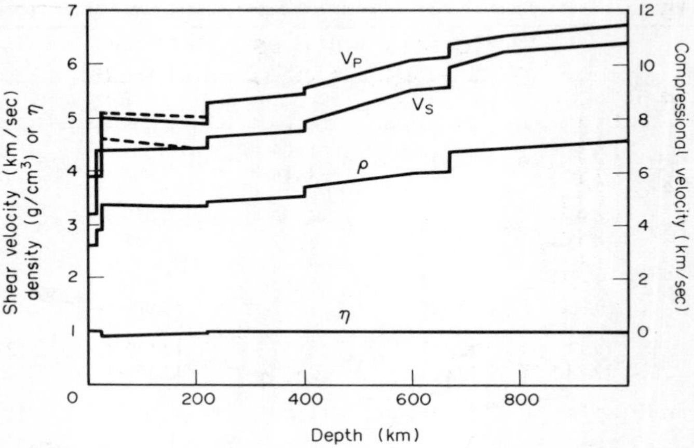

FIGURE 1.20 Upper mantle velocities, density and anisotropic parameter η in PREM. The dashed lines are the horizontal components of velocity. The solid curves are η, ρ and the vertical, or radial, components of velocity.

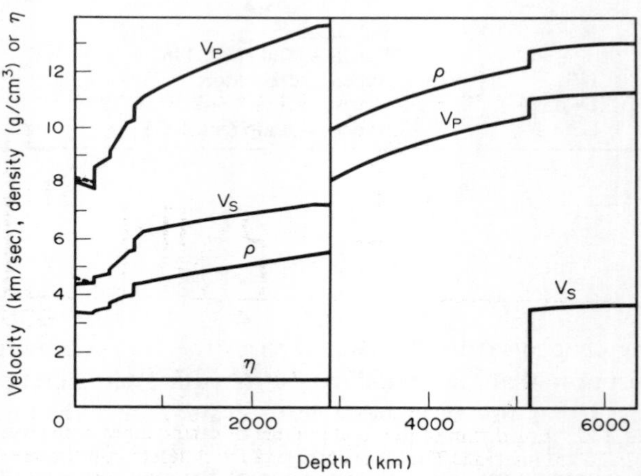

FIGURE 1.21 The PREM model. Dashed lines are the horizontal components of velocity. Where η is 1 the model is isotropic. The core is isotropic.

All these models are usually designated as SNREI models, which means "S pherical, N on-R otating, E lastic, I sotropic". Some models, B_1 for example, are built with the constraint to fit the Adams Williamson relation in the lower mantle and core.

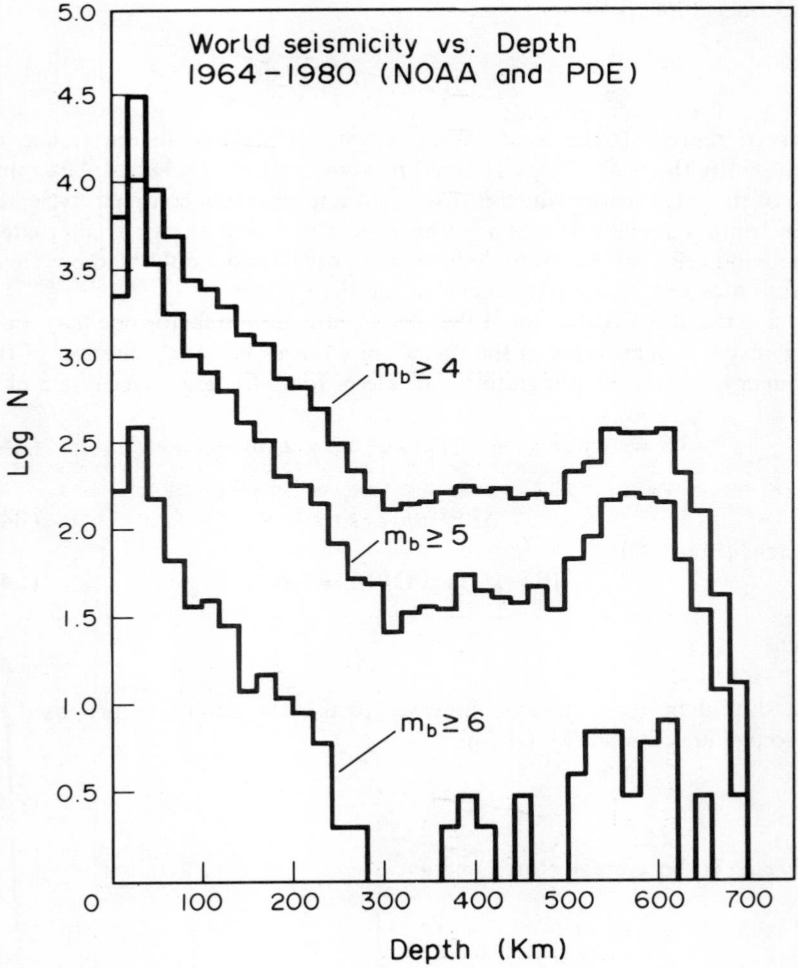

FIGURE 1.22 Logarithm of the total number of earthquakes in the world versus depth, in 20 km intervals. The three curves are for different cutoff magnitudes m_b, the one-second body wave magnitude (according to Vassiliou, Hager and Raefsky, 1984).

The Quality Factor Q

The PREM models are given for reference periods of 1 second and 200 seconds. They are the first models which include dissipation effects through the consideration of the quality factor Q. This parameter is usually defined by its inverse Q^{-1} called the specific attenuation factor or factor of anelasticity

(W. Kaula, 1968, p. 96).

$$Q^{-1} = \frac{\Delta E}{2\pi E} = \frac{1}{2\pi E} \oint \frac{dE}{dt}\, dt, \tag{1.42}$$

where E represents the total energy which is circulated in the system or equivalently the peak energy attained by some part of it, whereas ΔE is that part of the total energy which is dissipated into heat in a complete cycle, the cycle being a single oscillation in the system in which energy is alternately stored and released. An example of such an oscillation would be the cycle of vibration of one of the frequencies in the tidal waves.

Using the above definition of the specific attenuation factor one may quite easily express Q in terms of the phase lag ε for an harmonic motion of the frequency ω_i. Indeed, integrating the rate of loss of energy over one cycle,

$$\Delta E = \oint p\, de = -\oint p_0\, e_0\, \omega_i \cos \omega_i t \sin(\omega_i t + \varepsilon)\, dt, \tag{1.43}$$

hence

$$\Delta E = \pi p_0 e_0 \sin \varepsilon \tag{1.44}$$

whereas (Fig. 1.23)

$$W = \text{surface OPM} = \tfrac{1}{2} p_0 e_0 \tag{1.45}$$

Then

$$Q^{-1} = \sin \varepsilon \tag{1.46}$$

It should be recommended however to use the definition proposed by O'Connel and Budiansky (1978):

$$Q^{-1} = \frac{\Delta E}{4\pi E_a} \tag{1.45}$$

where E_a is the average stored energy.

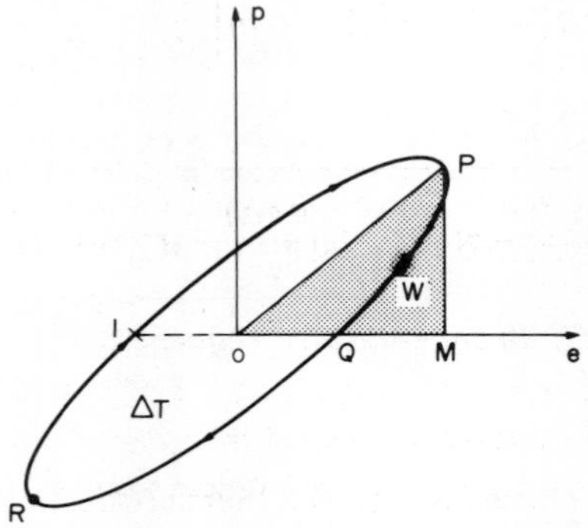

FIGURE 1.23 Comparison of the hysteresis loop with the peak energy represented by the surface OPM.

This allows one to relate the definition directly to the real and imaginary parts of the complex inelastic modulus of the deformation as well as of the stress (compliance)

$$Q = \frac{M_1}{M_2} = \frac{J_1}{J_2} \tag{1.46}$$

($M(\omega) = M_1(\omega) + iM_2(\omega)$ for the deformations; $J(\omega) = J_1(\omega) - iJ_2(\omega)$ for the stress). But all these definitions are equivalent for low values of Q^{-1}.

1.5 Stratification of the liquid core

The Brunt–Väisälä frequency

This important parameter aims to describe the role of the so-called *buoyancy force* which is nothing other than the Archimedean force. Let us suppose that, due to a small perturbation, a small element of the fluid moves from level r to level $r + \xi (\xi = dr)$. The difference of hydrostatic pressure between these two levels is

$$dp = -\rho g \xi. \tag{1.47}$$

Due to the adiabatic dilatation, according to eqn. (1.38), the density inside the element will change by:

$$(d\rho)_i = \rho \frac{dp}{k} = -\frac{\rho^2 g \xi}{k} \tag{1.48}$$

However, at new level, the density outside the element is changed by

$$(d\rho)_e = \xi \frac{d\rho}{dr} \tag{1.49}$$

The particle is thus submitted to a *restoring force*:

$$g[(d\rho)_e - (d\rho)_i] = \xi g \left[\frac{d\rho}{dr} + \frac{\rho^2 g}{k} \right] \tag{1.50}$$

Let us put:

$$\boxed{N^2(r) = -g(r)\rho^{-1}(r)\frac{d\rho(r)}{dr} - g^2(r)\rho(r)k^{-1}(r)} \tag{1.51}$$

$N(r)$ is called the *local Brunt–Väisälä frequency* (or sometimes "buoyancy frequency"). We have:

$$g[(d\rho)_e - (d\rho)_i] = -\rho \xi N^2 \tag{1.52}$$

and if we equate this expression to the inertial reaction of the particle:

$$\rho \frac{d^2 \xi}{dt^2} = -\rho \xi N^2$$

or:

$$\boxed{\frac{d^2\xi}{dt^2} + N^2\xi = 0} \tag{1.53}$$

One immediately sees that

if $N^2(r) > 0$ the stratification is stable: a particle, if displaced, comes back to its original position oscillating about it.

if $N^2(r) < 0$ the stratification is unstable: a particle, if displaced, continues its movement away from its original position.

while

if $N^2(r) = 0$ the stratification is neutral: a particle, if displaced remains a. its new position.

A stable stratification inhibits radial movements but not toroidal modes. In the case of an unstable stratification poloidal modes can be excited by the Archimedes force (buoyancy effects). The neutral stratification means that the Adams Williamson equation is satisfied. If this is the case for a proposed Earth core model, the core is said to be an *Adams Williamson core*. This may happen when the convection is so slow (or the conduction rapid enough) that fluid parcels adjust to their new environment when they move *radially*.

Despite the fact that seismic data do not directly reflect the $N^2(r)$ distribution, it is not difficult to calculate this basic parameter for any model Earth, and it is rather astonishing that it was never done until M. Smith (1976) and more recently Crossley and Rochester (1980)—see also Masters (1979) (see figure 1.25)—. The weak point in this calculation rests of course upon a correct estimation for $d\rho/dr$.

Obviously the terms in equation (1.45) more or less balance each other and the model of the Earth's interior constructed by Dziewonski, Hales and Lapwood (1975) on the basis of 1064 normal modes and 246 propagation times satisfies the Adams Williamson equation everywhere inside the Earth with a maximum deviation of 0.2% below 670 km. This should imply, of course, that deviations with respect to adiabaticity and to mineralogical homogeneity are very small. Some authors are of the opinion that N^2 is positive in the core except close to the inner core boundary where it is weakly negative.

Other expressions for the Brunt–Väisälä frequency

Pekeris and Accad (1972) have introduced a "stability parameter" $\beta(r)$ which is related to the Brunt–Väisälä frequency by the following equation:

$$\beta(r) = -k(r)\rho^{-1}(r)g^{-2}(r)N^2(r) \tag{1.54}$$

so that

$\beta > 0$ corresponds to instability

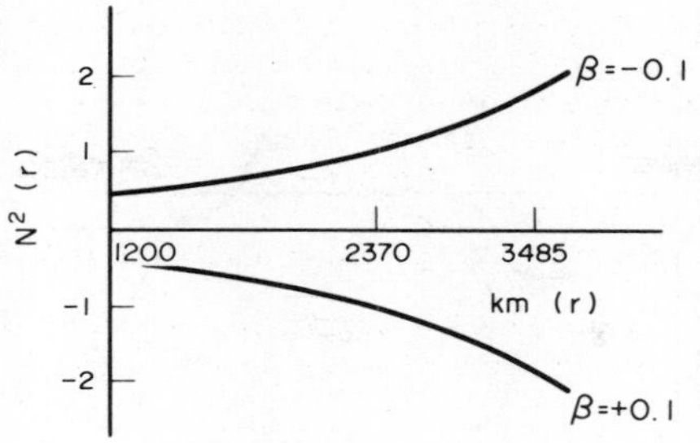

FIGURE 1.24 Computation of N^2 for typical values of β: $+0.1$ and -0.1.

$\beta < 0$ corresponds to stability
and $\beta = 0$ corresponds to the neutral stratification.

The Schwarzschild discriminant, used in astrophysics, is

$$A = -\frac{N^2}{g} \tag{1.55}$$

Figure 1.24 gives two examples of the correspondence between $\beta(r)$ and $N(r)$ easily calculated:

	r	$k(10^{12})$	g	ρ	$N^2(10^{-7})$
outer core boundary:	3485 km	6.8	1050	9.9	1.6
	2370 km	10.1	800	11.3	0.9
inner core boundary:	1200 km	12.0	600	11.9	0.36

Bullen had introduced still another parameter, deriving it from the Adams Williamson equation as follows:

$$\eta = -\frac{\phi}{g\rho}\frac{d\rho}{dr} = -\frac{k}{g\rho^2}\frac{d\rho}{dr} \tag{1.56}$$

η is simply related to the other parameters by:

$$\beta = 1 - \eta \qquad N^2 = \frac{g^2\rho}{k}(\eta - 1) \tag{1.57}$$

so that

$\eta < 1$ corresponds to instability
$\eta > 1$ corresponds to stability
and $\eta = 1$ corresponds to the neutral stratification.

However the most important is the relation between the Brunt–Väisälä frequency, the adiabatic gradient of temperature and the true gradient of temperature.

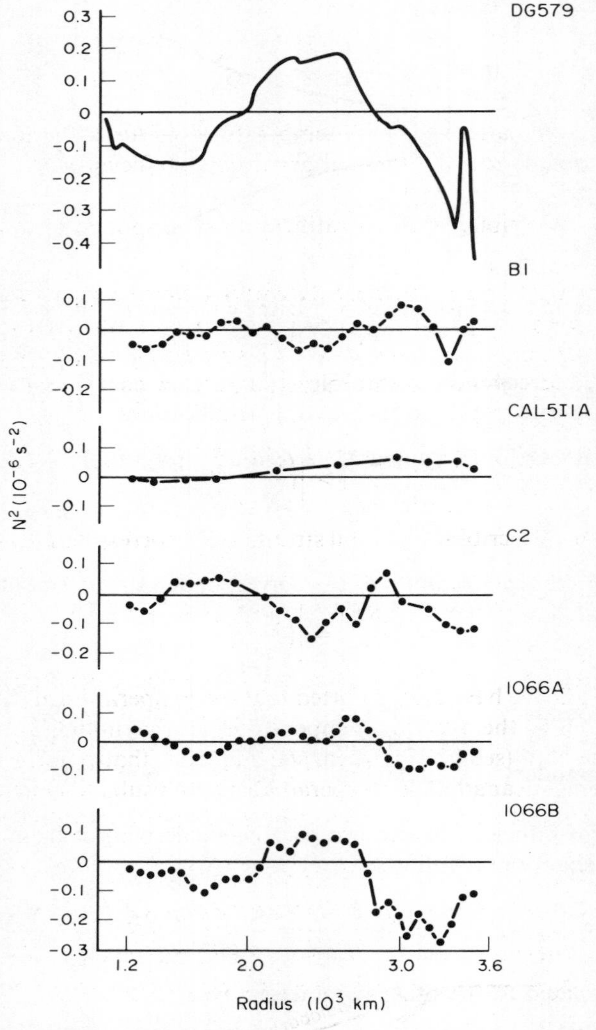

FIGURE 1.25 Brunt–Väisälä frequency profile across the core according to model DG579 (above) used by M. L. Smith (1976) and according to Crossley and Rochester (1980) computations for the other five models (see Table 1.6).

An excess of temperature over its adiabatic gradient would be responsible for an abnormal decrease of density. We can thus write (see eqn. 2.26)

$$\Delta\rho = -\alpha\rho\Delta T \text{ or } (d\rho)_e - (d\rho)_i = -\alpha\rho\left\{\left|\left(\frac{\partial T}{\partial r}\right)_s\right| - \left|\frac{\partial T}{\partial r}\right|\right\} dr \qquad (1.58)$$

and, from (1.52) we obtain

$$N^2(r) = \alpha(r)g(r)\left\{\left|\left(\frac{\partial T}{\partial r}\right)_s\right| - \left|\frac{\partial T}{\partial r}\right|\right\} \therefore \frac{\alpha g \Delta T}{L} \tag{1.59}$$

$(\partial T/\partial r)_s$ is the adiabatic gradient of temperature, $(\partial T/\partial r)$ is the actual gradient of temperature and α is the thermal expansion coefficient.
It results that:

$N^2(r) > 0$ describing stable stratification corresponds to a *subadiabatic gradient*

$$\left|\frac{dT}{dr}\right| < \left|\left(\frac{dT}{dr}\right)_s\right| \tag{1.60}$$

$N^2(r) < 0$ describing an unstable stratification corresponds to a *super-adiabatic gradient*

$$\left|\frac{dT}{dr}\right| > \left|\left(\frac{\partial T}{\partial r}\right)_s\right| \tag{1.61}$$

and $N^2(r) = 0$ describing a neutral stratification corresponds to an *adiabatic core* (see §2.11):

$$\left|\frac{dT}{dr}\right| = \left|\left(\frac{dT}{dr}\right)_s\right| \tag{1.62}$$

On figure 1.26 we have first admitted that the temperature at the inner core boundary is just the freezing temperature of the liquid mixture which composes the core (see §2.10). Then, starting from that point, we construct, across the core, linear adiabatic, superadiabatic and subadiabatic gradients of temperature.

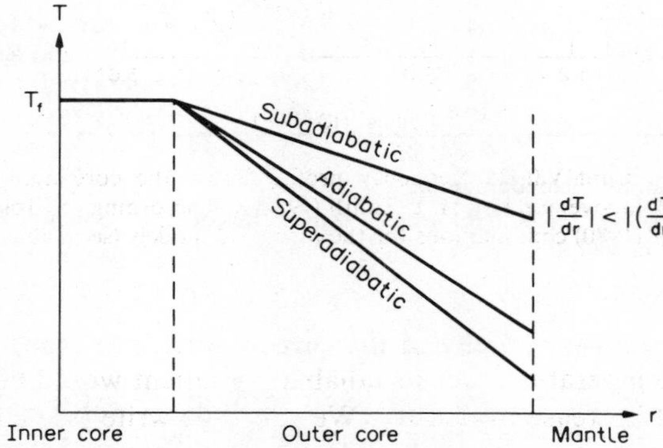

FIGURE 1.26 Profile of temperature gradients across the liquid outer core of the Earth. (T_f is the freezing temperature of the mixture).

It clearly appears that, in a case where the true gradient is superadiabatic the material just below the core mantle boundary should be colder than normal, it thus should go down, which should destabilize the core.

On the contrary, if the true gradient were subadiabatic the material just below the core mantle boundary should be warmer than normal and should stay where it is, keeping the core stable. However if a stable stratification ($N^2 > 0$, $\beta < 0$, $\eta > 1$) inhibits any poloidal convection, it does not affect a toroidal convection.

Finally let us point out that the long-term effect of convection is to reduce $|dT/dr|$ to its adiabatic value, and this will tend, of course, to stop the convection until the gradient has risen again. To maintain a *permanent* convection a small supply of heat is thus needed.

The core is strongly stably stratified when $A = N^2/4\Omega^2 > 1$. It is weakly stratified or in a state of "subcritical" stability when $A = N^2/4\Omega^2 < 1$. It is obvious that for a weaker stability, when the restoring Archimedean force is weaker, the period of oscillation will be longer, and subject to a stronger influence of the Coriolis force: inertial effects will become very important in that case. (see page 120, eqn 3.184).

We see here that the parameter N^2 has a more fundamental meaning than β or η but η has the advantage of algebraic simplicity. The advantage of the hydrodynamical formulation that we will adopt here, with respect to the Molodensky and Alterman formulation, is that N^2 appears explicitly in the equations.

TABLE 1.7 *Brunt–Väisälä frequency in different models of the Earth's core*

$N^2 (s^{-2})$	$N (s^{-1})$	$T = 2\pi/N$ (hours)	$A = N^2/4\Omega^2$	Reference
3.61×10^{-7}	6.01×10^{-4}	2.9	16.9	Smylie (1973)
3.38	5.81	3	16.1	Smith (1976)
1.90	4.36	4	9.0	Kennedy and Higgins (1973)
0.85	2.91	6	4.0	Crossley and Rochester (1980)
0.47	2.17	8	2.2(*)	Olson (1977)
0.081	0.9	19.4	0.38	Wahr (1981)

($4\Omega^2 = 0.213 \times 10^{-7}$)

A is the inverse rotational Froude number

* Olson has used a form $N^2 = \bar{N}^2 (r/b)^2$ where b is the radius of the core–mantle boundary so that $(r/b)^2$ varies from 0.12 to 1.0.

1.6 Topography of the core–mantle interface

If the hydrostatic equilibrium was perfectly realized inside the Earth, we could use the Clairaut equation to determine the flattening of surfaces of equal potential, pressure and density (grad $p = \rho$ grad V gives grad $V \wedge$ grad $\rho = 0$ by applying the curl operator to it). Denis and Ibrahim (1981) have obtained with the Dziewonski–Hales–Lapwood model (1975) a flattening of the core equal to 1/390.3.

However we know that the hydrostatic equilibrium is not perfectly realized (there are odd terms in the gravitational potential and the second-order term is slightly too large).

Garland (1957) suggested that the lower harmonics of the gravity field ($n = 3, 4, 5$) could be easily explained if the core–mantle boundary was not perfectly smooth and should have bumps and valleys of small height but of large surface extension. These low harmonics of the gravity field have very long wavelengths (13 000–7 000 km at the equator for the terms corresponding to $n = 3, 4, 5, 6$) so that it seems impossible to ascribe them to shallow density inequalities, even up to depths of 400 km and 650 km where discontinuities exist but with only small density jumps of about 0.3 g cm^{-3}. On the contrary the density jump at the core–mantle interface is as large as 4.5 g cm^{-3}. In 1968 Hide and Horai showed that bumps of several hundred metres height could very well explain the gravimetric observations.

As we will see later (§3.5) bumps (or valleys) of the boundary would have a considerable influence on the hydrodynamic convection which can exist in the form of Taylor columns inside the core (figure 3.3).

An evaluation of the size of such hypothetical bumps can easily be made by admitting that the lower orders in the Standard Earth Gravitational Potential as described by eqn. 1.4 should be attributed to such irregularities.

Let us represent the core–mantle interface by a very thin layer of surface density:

$$\sigma(\theta, \lambda) = \rho' \sum_n \sum_m (\varepsilon_{nm}^{(1)} \cos m\lambda + \varepsilon_{nm}^{(2)} \sin m\lambda)\, P_n^m(\cos\theta) \tag{1.63}$$

where $\varepsilon_{nm}^{(1)}$ and $\varepsilon_{nm}^{(2)}$ describe the shape of the core–mantle interface, and

$$\rho' = \rho_{\text{core}} - \rho_{\text{mantle}} \sim 4\ \text{g cm}^{-3}. \tag{1.64}$$

The gravitational potential exerted by this layer upon an external point P

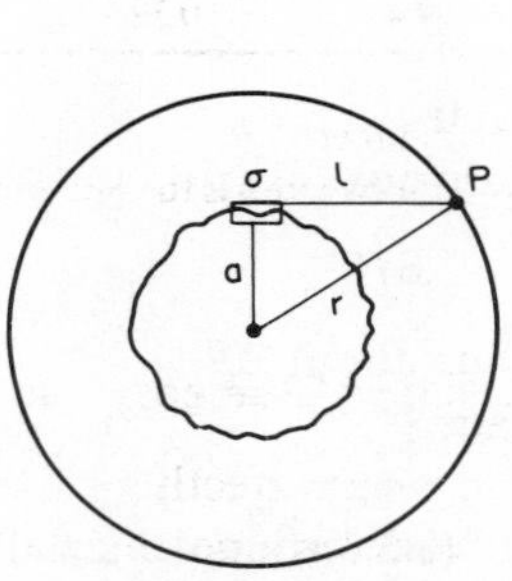

FIGURE 1.27.

located at the Earth's surface is:

$$V_E = G\iint \sigma(\theta,\lambda)\frac{ds}{l} = \sum_n\sum_m \frac{4\pi G\rho' a}{2n+1}\left(\frac{a}{r}\right)^{n+1}$$

$$\{\varepsilon_{nm}^{(1)}\cos m\lambda + \varepsilon_{nm}^{(2)}\sin m\lambda\}\, P_n^m, \quad (1.65)$$

as

$$\frac{4\pi G\bar{\rho}a}{3} = \bar{g} \quad \text{with} \left(\frac{a}{r}\right) \sim 0.5, \quad \bar{\rho} = 5.52 \quad (1.66)$$

we have:

$$\frac{V_E}{\bar{g}} = \sum_n\sum_m \frac{3\rho'}{\bar{\rho}}\frac{(0.5)^{n+1}}{2n+1}(\varepsilon_{nm}^{(1)}\cos m\lambda + \varepsilon_{nm}^{(2)}\sin m\lambda)\,P_n^m(\cos\theta) \quad (1.67)$$

Let us take, as an example, the term of order (5, 1) in the geopotential:

$$\bar{C}_{51} = 5.38\times10^{-8}, \quad \bar{S}_{51} = 9.79\times10^{-8}$$

These coefficients are normalized according to the formula:

$$C_{nm} = \sqrt{\frac{(2n+1)(n-m)!}{2(n+m)!}}\,\bar{C}_{nm} \quad (1.68)$$

Thus, for $n = 5$, $m = 1$, this square root is 0.428,

$$C_{51} = -2.3\times10^{-8} \quad S_{51} = -4.2\times10^{-8}$$

and

$$V_5^1 = -\frac{GM}{a}\left(\frac{a}{r}\right)^6 [C_{51}\cos\lambda + S_{51}\sin\lambda]\,P_5^1(\cos\theta)$$

and at $r = a$, it gives

$$V_5^1(r=a) = -a\bar{g}\,[C_{51}\cos\lambda + S_{51}\sin\lambda]\,P_5^1(\cos\theta)$$

with

$$P_5'(\cos\theta) = -\frac{15}{8}(21\cos^4\theta - 14\cos^2\theta + 1)\sin\theta$$

Now, at $\theta = 0°$: $P_{51} = 0$

$\theta = 45°$: $P_{51} = +0.994$

$\theta = 90°$: $P_{51} = -1.875$

its maximum variation is thus about 2.9.

Thus, at the Earth's surface:

$$\Delta a = \frac{\Delta V}{\bar{g}} \approx [6.37\times10^6(2.3\cos\lambda + 4.2\sin\lambda)\times10^{-8}\times2.9]\text{ metres}$$

$$= [0.42\cos\lambda + 0.77\sin\lambda]\text{ metres} \quad (1.69)$$

Equating the coefficients of eqns (1.67) and (1.69) it results that

$$\varepsilon_{nm}^{(1)} = \left| \frac{3 \times 4}{5{,}52} \frac{(0.5)^6}{11} \right|^{-1} 0.42 \simeq 140 \text{ metres}$$

and

$$\varepsilon_{nm}^{(2)} = (0.003)^{-1}\, 0.77 \simeq 260 \text{ metres}$$

which means that a bump of some 200 metres at the core–mantle interface can be responsible for the observed C_{51}, S_{51} terms in the geopotential.

A seismic path grazing the core–mantle boundary can be observed when the angular distance between the seismic source and the receiving seismographic station is around 95°.

This may allow to sample the core–mantle region. Perturbations observed on some registrations can indeed be ascribed either to heterogeneity in the lowermost mantle layer (called D″ layer), either to such bumps of the boundary with few hundred meters height (Doornbos, 1978, Ruff and Helmberger, 1982).

We will return to this problem after having developed the theory of boundary layers (see §5.4).

Bibliography

Adams, R. D. (1972) Multiple core reflections from a Novaya Zemlya explosion. *Bull. Seism. Soc. Am.*, **62,** 1063–71.

Alterman, Z., Jarosch, H. and Pekeris, C. L. (1959) Oscillations of the Earth. *Proc. R. Soc. Lond.*, **A252,** 80–95.

Anderson, O. L. (1986) Properties of iron at the Earth's core conditions. Geophys. *J. R. Astr. Soc.* **84**, 561–579.

Barnes, R. T. H., Hide, R., White, A. A. and Wilson, C. A. (1983) Atmospheric angular momentum fluctuations length of the day changes and polar motion. *Proc. R. Soc. Lond.*, **A387,** 31–73.

Bolt, B. A. and Niazi, M. (1984) S Velocities in D″ from diffracted SH-waves at the core boundary. *Geophys. J. R. Astr. Soc.* **79**, 825–34.

Bolt, B. A. and Qamar, A. (1972) Observations of pseudo-aftershocks from underground nuclear explosions. *Phys. Earth Plan. Int.*, **5,** 400–2.

Bolt, B. A. and Uhrhammer, R. (1975) Resolution techniques for density and heterogeneity in the Earth. *Geophys. J.R.Astr. Soc.*, **42,** 419–35.

Brush, S. G. (1980) Discovery of the Earth's core. *Am. J. Phys.*, **48,** 705–24.

Buchbinder, G. G. R. (1972) Travel times and velocities in the outer core from PnKP. *Earth Plan. Sci. Lett.*, **14,** 161–8.

Bukowinski, M. S. T. (1976) On the electronic structure of iron at core pressures. *Phys. Earth Plan. Int.*, **13,** 57–66.

Bullen, K. E. (1965) Models for the density and elasticity of the Earth's lower core. *Geophys. J. R. Astr. Soc.*, **9,** 233.

Bullen, K. E. (1967) Note on the coefficient eta. *Geophys. J. R. Astr. Soc.*, **13,** 459.

Bullen, K. E. and Haddon, R. A. W. (1973) Some recent work on Earth models, with special reference to core structure. *Geophys. J. R. Astr. Soc.*, **34,** 31–8.

Carter, W. E., Robertson, D. S. and MacKay, J. R. (1985) Geodetic Radio Interferometric Surveying: Applications and Results. *J. Geophys. Res.* **90**, 4577–88.

Chandler, S. C. (1891) On the variation of latitude. *Astron. J.*, **XI**(249), 65–70.

Crossley, D. J. and Rochester, M. G. (1980) Simple core undertones. *Geophys. J. R. Astr. Soc.*, **60,** 129–61.

Denis, C. (1974) Oscillations de configurations sphériques autogravitantes et applications à la terre. Univ. de Liege, thèse de doctorat en sci., 2 vols.
Denis, C. and Ibrahim, A. (1980) Programme numérique permettant de représenter des modèles terrestres planétaires et stellaires de manière cohérente. *Bull. Inform. Marées Terrestres.* **83,** 5236–93.
Denis, C. and Ibrahim A. (1981) On a self-consistent representation of Earth models, with an application to the computing of internal flattening. *Bull. Geod.*, **55,** 179–95.
Djurovic, D. (1976) Détermination du nombre de Love k et du facteur Λ affectant les observations du temps universel. *Astron. Astrophys.*, **47,** 325–332.
Doornbos, D. J. (1978) On seismic-wave scattering by a rough core–mantle boundary. *Geophys. J. R. Astr. Soc.* **53,** 643–62.
Dziewonski, A. M. and Anderson, D. L. (1981) Preliminary reference Earth model. *Phys. Earth Plan. Int.*, **25,** 297–356.
Dziewonski, A. M. and Gilbert, F. (1983) Observations of normal modes from 84 recordings of the Alaskan earthquake of 28 march 1974; spheroidal overtones. *Geophys. J. R. Astr. Soc.*, **35,** 401–37.
Dziewonski, A. M., Hales, A. L. and Lapwood, E. R. (1975) Parametrically simple models consistent with geophysical data. *Phys. Earth Plan. Int.*, **10,** 12–48.
Elsasser, W. M. and Isenberg, I. (1949) Electronic phase transition in iron at extreme pressure. *Phys. Rev.*, **76,** 469A.
Friedlander, S. (1985) Internal oscillations in the Earth's fluid core. *Geoph. J. R. Astr. Soc.*, **80,** 345–61.
Gans, R. F. (1972) Viscosity of the Earth's core. *J. Geophys. Res.*, **77,** 360–6.
Gardiner, R. B. and Stacey, F. D. (1971) Electrical resistivity of the core. *Phys. Earth Plan. Int.*, **4,** 406–18.
Garland, G. D. (1957) The figure of the earth's core and the non-dipole field. *J. Geophys. Res.*, **66,** 486–7.
Gilbert, F. and Dziewonski, A. M. (1975) An application of normal mode theory to the retrieval of structural parameters and source mechanisms from seismic spectra. *Phil. Trans. R. Soc. Lond.*, **278** (A1280), 187.
Gubbins, D. (1982) Stable regions in the Earth's liquid core. *Geophys. J. R. Astr. Soc.*, **68,** 241–51.
Hide, R. and Horai, K. (1968) On the topography of the core-mantle interface. *Phys. Earth Plan. Int.*, **1,** 305–8.
Jacobs, J. A. (1975) The Earth's Core. Academic press.
Jeanloz, R. (1979) Properties of iron at high pressures and the state of the core. *J. Geophys. Res.*, **84,** 6059–69.
Jordan, T. H. and Anderson, D. L. (1974) Earth structure from free oscillations and travel. *Geophys. J.R. Astr. Soc.*, **36,** 411–59.
Kaula, W. (1968) An Introduction to Planetary Physics. John Wiley and Sons.
Kennedy, G. C. and Higgins, G. H. (1973) The core paradox. *J. Geophys. Res.*, **78**(5), 900–3.
Masters, T. G. (1979) Observational constraints on the chemical and thermal structure of the Earth's deep interior. *Geophys. J. R. Astr. Soc.*, **57,** 507–34.
McLeod, M. G. (1985) On the geomagnetic jerk of 1969. *J. Geophys. Res.*, **90,** 4597–610.
Melchior, P. (1971–3) Physique et Dynamique Planetaires. (1) *Geodesie et Astronomie Geodesique*; (2) *Gravimetrie* (1971) (3) *Geodynamique* (1972) (4) *Geodynamique* (1973). Vander, Bruxelles.
Melchior, P. (1971) Precession-nutations and tidal potential. *Celest. Mech.*, **4,** 190–212.
Melchior, P. (1980) Luni solar nutation tables and the liquid core of the Earth. *Astron. Astrophys.*, **87,** 365–8.
Melchior, P. (1982) *The Tides of the Planet Earth*, 2nd edn. Pergamon Press.
Molodensky, M. S. (1961) The theory of nutations and diurnal Earth tides. *Comm. Observ. Roy. Belg.*, **188**; *S. Geoph.* **58,** 25.
Newcomb, S. (1892) On the dynamics of the Earth's rotation with respect to periodic variations of latitude. *Monthly Not. R. Astr. Soc.*, **52,** 336–41.
O'Connell, R. J. and Budiansky, B. (1978) Measures of dissipation in viscoelastic media *Geophys. Res. Lett.*, **5,** 5–8.
Pekeris, C. L. and Accad, Y. (1972) Dynamics of the liquid core of the Earth. *Phil. Trans. R. Soc. Lond.*, **A273,** 237.
Poinsot, L. (1854) Theorie nouvelle de la rotation des corps. *Connaissance des Temps Additions*, 3–134.

Press, F. (1970) Earth models consistent with geophysical data. *Phys. Earth Plan. Int.*, **3**, 3–22.
Qamar, A. and Eisenberg, A. (1974) The damping of core waves. *J. Geophys. Res.*, **79**, 758–765.
Ruff, L. J. and Helmberger, D. V. (1982) The structure of the lowermost mantle determined by short-period P-wave amplitudes. *Geophys. J. R. Astr. Soc.* **68**, 95–119.
Saigey, J. (1829) *Petite Physique du Globe*, Tome II, P. 185. Cited in Tisserand's *Celestial Mechanics*, Vol. 2.
Smith, M. (1976) Translational inner core oscillations of a rotating slightly elliptic Earth. *J. Geophy. Res.*, **81**, 3055–65.
Spiliopoulos, S. and Stacey, F. D. (1984) The Earth's thermal profile: is there a mid-mantle thermal boundary layer? *J. Geodynam.*, **1**, 61–77.
Stiller, H., Franck, S. and Mohlmann, D. (1984) Geodynamics and state of the Earth's interior. *J. Geodynam.*, **1**, 79–100.
Stix, M. (1983) The rotation and the magnetic field of the Earth. *Tidal Friction and the Earth's Rotation*, II. Springer-Verlag, 89–116.
Vassiliou, M. S., Hager, B. H. and Raefsky, A. (1984) The distribution of earthquakes with depth and stress in subducting slabs. *J. Geodynam.*, **1**, 11–28.
Wahr, J. (1981) Body tides on a elliptical rotating elastic and oceanless Earth. *Geophs. J. R. Astr. Soc.*, **64**, 677–704.
Wright, C. (1973) Observations of the multiple core reflections of the PNKP and SNKP type and regional variations at the base of the mantle. *Earth Plan. Sci. Lett.*, **19**, 453–460.
Wright, C. (1973) Array studies of P phases and the structure of the D″ region of the mantle. *J. Geophys. Res.*, **78**, 4965–82.
Wright, C., Muirhead, K. J. and Dixon, A. E. (1985) The P wave velocity structure near the base of the mantle. *J. Geophys. Res.*, **90** (B1), 623–34.
Wunsch, C. (1974) Simple models of the deformation of the Earth with a fluid core. *Geophys. J. R. Astr. Soc.*, **39**, 413–19.
Yoder, C. F., Williams, J. G., Dickey, J. O., Schutz, B. E., Eanes, R. J. and Tapley, B. D. (1983) Secular variation of Earth's gravitational harmonic J2 from LAGEOS and the non-tidal acceleration of Earth rotation. *Nature*, **303**, 757.

CHAPTER 2

Thermodynamics

As shown in chapter 1, the pressure in the deep interior of the Earth can reach a few megabars. This causes sufficiently large changes in densities and in temperature, by adiabatic compression, to justify the application of thermodynamics to the Earth's interior as in the case of ideal gases.

2.1 Fundamental thermodynamic relations*

It is essential to point out that there are two kinds of variables in thermodynamics:

extensive (additive) variables:	V (specific volume), ↓	S (entropy), ↓	N (number of elements) ↓
and their conjugate *intensive* variables:	P (pressure),	T (temperature),	μ (chemical potential)

(specific volume is such as $\rho V = 1$, ρ being the density)

Conservation laws apply to extensive variables. Intensive variables which are also called "field variables" (velocity, stress tensor, density) can be expressed as $f(x_1, x_2, x_3, t)$.

Intensive variables are easier to measure experimentally. The *internal energy*, which is the energy accumulated inside the system by heating (dQ) and mechanical work (dW):

$$dE = dQ + dW = TdS - PdV \tag{2.1}$$

depends on S and V.

The *thermodynamic potentials*,† which are of convenient use because they include the external permanent interactions of the surroundings, are:

Enthalpy	$H = E + PV$	(2.2)
Free Helmholtz energy	$F = E - TS$	(2.3)
Free Gibbs energy	$G = E - TS + PV$	(2.4)

* Our basic reference is here Landau and Lifschitz (1967).

† S is expressed in joules $kg^{-1} K^{-1}$ or in joules $mole^{-1} K^{-1}$ or in calories $mole^{-1}$ (1 calorie = 4.1846 joule; 1 eV ~ 1.75×10^{-12} erg).

Enthalpy permits writing the first principle of thermodynamics without explicitly introducing the work done by pressure forces.

These potentials are all additive, thus depending upon the number of elements N. From (2.1) it follows that in a closed system:

$$dH = TdS + VdP \tag{2.5}$$

$$dF = -SdT - PdV \tag{2.6}$$

$$dG = -SdT + VdP \tag{2.7}$$

where one immediately observes that *only G depends upon intensive variables* (T, P) *only*, while it is itself proportional to the number of particles (extensive variable). Thus we may write:

$$G = Nf(T, P) \tag{2.8}$$

while

$$\left.\begin{aligned} E &= Nf'\left(\frac{S}{N}, \frac{V}{N}\right) \\ F &= Nf''\left(\frac{V}{N}, T\right) \\ H &= Nf'''\left(\frac{S}{N}, P\right) \end{aligned}\right\} \tag{2.9}$$

Consequently we can put:

$$\mu = \left(\frac{\partial G}{\partial N}\right)_{T,P} = \frac{G}{N} \tag{2.10}$$

Because the other potentials and E are also extensive quantities we still can write

$$\mu = \left(\frac{\partial E}{\partial N}\right)_{S,V} = \left(\frac{\partial H}{\partial N}\right)_{S,P} = \left(\frac{\partial F}{\partial N}\right)_{T,V} = \left(\frac{\partial G}{\partial N}\right)_{T,P} \tag{2.11}$$

so that we have to conclude that in an open system:

$$dE = TdS - PdV + \mu dN \tag{2.12}$$

$$dH = TdS + VdP + \mu dN \tag{2.13}$$

$$dF = -SdT - PdV + \mu dN \tag{2.14}$$

$$dG = -SdT + VdP + \mu dN \tag{2.15}$$

μ is called the *chemical potential*. These relations are important when we will have to deal with liquid mixtures like in the Earth's core. Several phenomenological state equations can now be written as:

$$P(V, T) = -\left(\frac{\partial F}{\partial V}\right)_{T,N} = -\left(\frac{\partial E}{\partial V}\right)_{S,N} \tag{2.16}$$

$$S(V, T) = -\left(\frac{\partial F}{\partial T}\right)_{V,N} = -\left(\frac{\partial G}{\partial T}\right)_{P,N} \tag{2.17}$$

$$V(T, P) = \left(\frac{\partial G}{\partial P}\right)_{T,N} = \left(\frac{\partial H}{\partial P}\right)_{S,N} \tag{2.18}$$

etc . . .

As V is the specific volume equation (2.13) can also be written

$$\text{grad } P = \rho \text{ grad } H - \rho T \text{ grad } S - \rho\mu \text{ grad } N \tag{2.19}$$

2.2 Definitions of fundamental thermodynamic parameters

Specific heats

Specific heat is defined as the heat necessary to apply to 1 g of matter to raise its temperature by 1°. It is defined under two different conditions:

at constant volume

$$C_V = \left(\frac{\partial E}{\partial T}\right)_V = T\left(\frac{\partial S}{\partial T}\right)_V \tag{2.20}$$

and at constant pressure

$$C_P = \left(\frac{\partial H}{\partial T}\right)_P = T\left(\frac{\partial S}{\partial T}\right)_P \tag{2.21}$$

in this second case there is a volume variation to counteract so that $C_P > C_V$: $C_P - C_V = R \sim 2$ calories g^{-1} K^{-1} (Mayer's law, R: constant of perfect gas). Moreover as $E = 3NkT$:

$$C_V = 3Nk \tag{2.22}$$

(N: Avogadro number, k: Boltzmann constant, T: absolute temperature).

For monoatomic gas:

$$C_P/C_V = 5/3 = 1.67 \tag{2.23}$$

Compressibilities

These are defined as follows:

at constant temperature:

$$\beta_T = \frac{1}{k_T} = -\frac{1}{V}\left(\frac{\partial V}{\partial P}\right)_T = -\rho\left(\frac{\partial 1/\rho}{\partial P}\right)_T = +\frac{1}{\rho}\left(\frac{\partial \rho}{\partial P}\right)_T \tag{2.24}$$

and at constant entropy (adiabatic):

$$\beta_S = \frac{1}{k_S} = -\frac{1}{V}\left(\frac{\partial V}{\partial P}\right)_S \tag{2.25}$$

k_T and k_S are the respective *incompressibility moduli* and have the dimensions of a pressure; k_S is often called *bulk modulus* in seismology.

The thermal expansion coefficient (*cubic dilatation*)

This is:

$$\alpha = +\frac{1}{V}\left(\frac{\partial V}{\partial T}\right)_P = -\frac{1}{V}\left(\frac{\partial S}{\partial P}\right)_T = -\frac{1}{\rho}\left(\frac{\partial \rho}{\partial T}\right)_P \tag{2.26}$$

by using equations (2.17) and (2.18).

Fundamental relation between specific heats and compressibilities

By Jacobian manipulations one easily gets

$$\left(\frac{\partial V}{\partial P}\right)_S = \frac{\partial(V,S)}{\partial(P,S)} = \frac{\partial(V,S)/\partial(V,T)}{\partial(P,S)/\partial(P,T)}\cdot\frac{\partial(V,T)}{\partial(P,T)} = \frac{(\partial S/\partial T)_V}{(\partial S/\partial T)_P}\cdot\left(\frac{\partial V}{\partial P}\right)_T$$
$$= \frac{C_V}{C_P}\left(\frac{\partial V}{\partial P}\right)_T \tag{2.27}$$

thus

$$\frac{C_V}{k_T} = \frac{C_P}{k_S} \tag{2.28}$$

2.3 Adiabatic gradient of temperature

The word "isentropic" is also used for "adiabatic".
One may write, by using Jacobians again:

$$\left(\frac{\partial T}{\partial P}\right)_S = \frac{\partial(T,S)}{\partial(P,S)} = \frac{\partial(T,S)/\partial(P,T)}{\partial(P,S)/\partial(P,T)} = -\frac{(\partial S/\partial P)_T}{(\partial S/\partial T)_P} \tag{2.29}$$

which, with

$$\left(\frac{\partial S}{\partial P}\right)_T = -\left(\frac{\partial V}{\partial T}\right)_P \tag{2.30}$$

and the definitions (2.21) and (2.26) give

$$\left(\frac{\partial T}{\partial P}\right)_S = \frac{T}{C_P}\left(\frac{\partial V}{\partial T}\right)_P = \frac{\alpha V T}{C_P} \tag{2.31}$$

Introducing the adiabatic hypothesis ($dS = 0$) and $dN = 0$ in (2.19), one has $dP/dr = -\rho g$ while $\rho V = 1$ (V being the specific volume) we obtain:

$$\left(\frac{\partial T}{\partial r}\right)_S = \left(\frac{\partial T}{\partial P}\right)_S \frac{dP}{dr} \tag{2.32}$$

or finally

$$\boxed{\left(\frac{\partial T}{\partial r}\right)_S = -\frac{\alpha g T}{C_P}} \tag{2.33}$$

an equation which allows us to evaluate the adiabatic gradient of temperature starting from a depth r where the temperature is known, provided that $\alpha(r)$, $C_P(r)$ and $g(r)$ are also known (see Stacey, 1977b).

2.4 The Wiedemann–Franz law

Thermal and electrical conductivities are transport properties which are related by a law called the Wiedemann–Franz law.

Thermal conductivity

Let us consider two vertical layers of metal each having a section S and a thickness equal to the mean free path λ of an electron. During the time $\tau = \lambda/u$ the number of electrons of velocity u, in transit from one layer to the other (in only one sense), is $N = (1/6)nS\lambda$ ($n = \Delta N/\Delta V$ being their concentration). They carry an energy $W_1 = N\varepsilon_1$ from left to right but also $W_2 = N\varepsilon_2$ from right to left. Thus

$$Q = W_1 - W_2 = N(\varepsilon_1 - \varepsilon_2) = \frac{1}{6}nS\lambda \cdot \frac{3}{2}k(T_1 - T_2)$$

and the heat flow is

$$\frac{\Delta Q}{S\Delta t} = \frac{1}{4}n\lambda k \frac{T_1 - T_2}{X_2 - X_1} \cdot \frac{X_2 - X_1}{\Delta t} \quad \text{with } X_2 - X_1 = 2\lambda = 2u\Delta t$$

we can put

$$\frac{\Delta Q}{S\Delta t} = -K\frac{\Delta T}{\Delta x} = -\frac{1}{2}n\lambda uk\frac{\Delta T}{\Delta x}$$

and the *thermal conductivity* is defined by

$$\boxed{K = \tfrac{1}{2}n\lambda uk} \tag{2.34}$$

(note that λu is the thermal diffusivity χ which is expressed in $cm^2\,s^{-1}$ or stokes).

Electrical conductivity

An electrical field of intensity E causes a migration of the charge carriers: an electron is indeed submitted to a force $F = eE$ (e being the charge of one electron) and moves with an acceleration $a = F/m = eE/m$ until it collides with an ion. Its mean speed in the conducting material is

$$\bar{v} = \frac{x - x_0}{\tau} = \frac{a\tau}{2}$$

τ being the mean free time or time of the mean free path. Thus:

$$\bar{v} = \frac{e\tau E}{2m}$$

The density of current across the section S with a concentration of free charges $n = \Delta N/\Delta V$ and $i = \Delta q/\Delta t = e\Delta N/\Delta t = en\Delta V/\Delta t = enS\bar{v}$ is

$$j = \frac{i}{S} = en\bar{v} = \frac{e^2 n\tau}{2m} E = \sigma E$$

where σ is the *electrical conductivity*:

$$\sigma = \frac{e^2 n\tau}{2m} = \frac{e^2 n\lambda}{2mu} \tag{2.35}$$

(as $\tau u = \lambda$ mean free path).

We can now take the ratio of both conductivities (with the kinetic energy $mu^2/2 = 3kT/2$):

$$\boxed{\frac{K}{\sigma} = \frac{kmu^2}{e^2} = \frac{3k^2 T}{e^2} = 2.23 \times 10^{-8}\,T \quad \text{Watt}\,\Omega\,\text{deg}^{-2}} \tag{2.36}$$

This is the Wiedemann–Franz law, and 2.23×10^{-8} is called the *universal Lorentz number*. It is valid for solids but liquids do not deviate too much, as is experimentally shown.

This law allows estimation of the thermal conductivity of a material if we know its electrical conductivity. It is, in principle, more feasible to determine the electrical conductivity inside the Earth than its thermal conductivity.

2.5 Measurements of electrical conductivity

The resistivity to current flow is measured in ohm metres, that is the resistance of a cubic metre of the material. The conductivity is the reciprocal of the resistivity, it is measured in *Siemens per metre* (S m^{-1} or mho m^{-1}). It can be determined if

1. there is an energy source on or above the Earth's surface, for example an alternating magnetic field excited by the solar activity (diurnal variation S_q);
2. measurements of a combination of magnetic and electrical field are made at stations over a broad region;

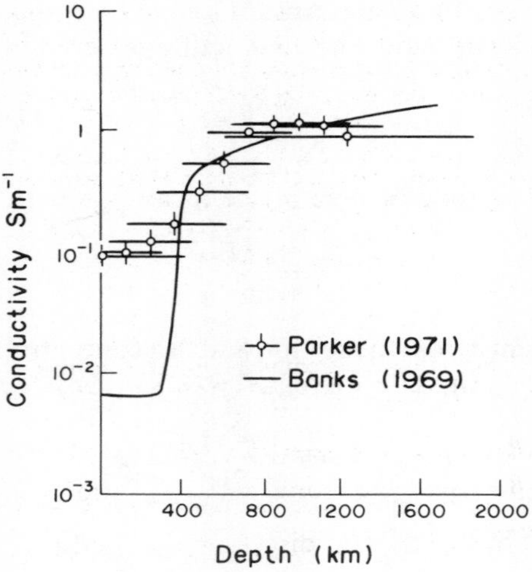

FIGURE 2.1 Profile of the electric conductivity in the upper mantle (Garland, 1981).

3. there is a mathematical procedure for inverting the observed field quantities into a model of underground conductivity distribution based upon Maxwell's equations.

Note that the depth of penetration or "*skin depth*" ($1/e$ of surface value, see §4.3) strongly depends upon the frequency. For a rock of resistivity $10^4\,\Omega$m this depth is about 50 km for $\nu = 1$ Hertz and 400 km for $\nu = 1/60$ Hertz (Garland, 1981).

The major jump at a depth of 400 km (figure 2.1) is related to the crystallographic change in olivine to spinel structure. Lateral variations of conductivity give information about composition, water, content, partial melting, etc. At shallower depths neither the highly resistive continental crust nor the conducting asthenosphere are resolved.* The conductivity of the deeper mantle is very uncertain because no external variations of sufficiently long periods are available. Estimates of the maximum possible conductivity have been made, based on the assumption that variations of a few years period, seen at magnetic observatories, are produced in the core and propagate through the mantle. This gives, for the upper mantle: $\sigma < 10^{-1}$ S m^{-1} and for the lower mantle: $\sigma \sim 10^2$ S m^{-1} (Smylie. 1965) or $\sigma \sim 150$ S m^{-1} as an upper bound (Achade *et al.*, 1980).

* The upper boundary of the lower crust can be defined as this interface where the electrical conductivity is the highest, and this may be due to its water content. This interface is called the Conrad discontinuity and is 13–15 km deep while the Mohorovicic discontinuity is about 30 km deep.

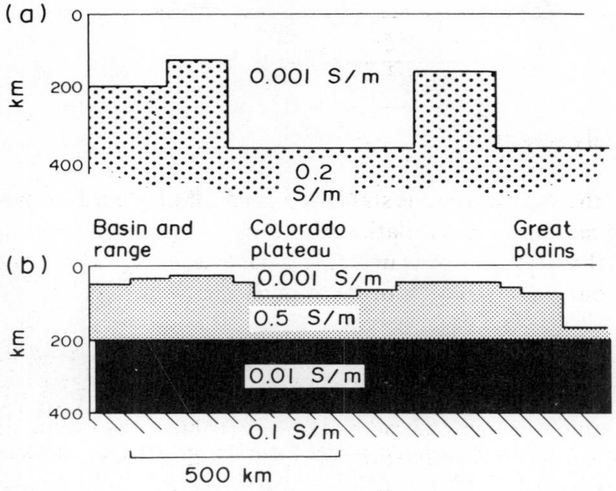

FIGURE 2.2 Two-dimensional conductivity models from GDS in the USA at 38°N; (a) and (b) are alternative models that fit the magnetic variation data, but (b) was also matched to other geophysical models (after Gough, 1973).

We have found, for example

in the upper mantle	$\sigma = 10^{-1}$ to $10^{-2}\,\mathrm{S\,m^{-1}}$	(*EOS* **63** (36), 769)
in the lower mantle	$\sigma = 10^{2}\,\mathrm{S\,m^{-1}}$	(Smylie, *Geophys. J. R. Astr. Soc.*, 1964, 169)
at the core boundary	$\sigma \sim 10^{3}\,\mathrm{S\,m^{-1}}$	(*KAPG*)

This low conductivity of the upper mantle is responsible for the screening of the toroidal geomagnetic field of the Earth's core.

According to Gardiner and Stacey (1971):

$$\text{in the core} \quad \sigma \sim 5.10^{5}\,\mathrm{S\,m^{-1}} \tag{2.37}$$

while by extrapolating for temperature, pressure and compressibility laboratory measurements made with the alloy $(Fe_{90}Ni_{10})_3\,S_2$ at 100°C, Johnston and Strens (1973) propose, at 1000 °C and zero pressure:

$$\sigma = 2.7 \cdot 10^{5}\,\mathrm{S\,m^{-1}} \tag{2.38}$$

This is not very different from the preceding one, and in any case is a factor of 1000 or more respective to the lower mantle. Note, however, that at the interface the discontinuity should not exceed one order of magnitude.

The probability of a very low conductivity of the mantle is corroborated by a result obtained by Hide and Malin (1981), who succeeded in calculating an "electrical radius" of the core on the hypothesis that the mantle has zero conductivity: $r = 3450$ km, close to the "seismological radius" $r = 3485$ km.

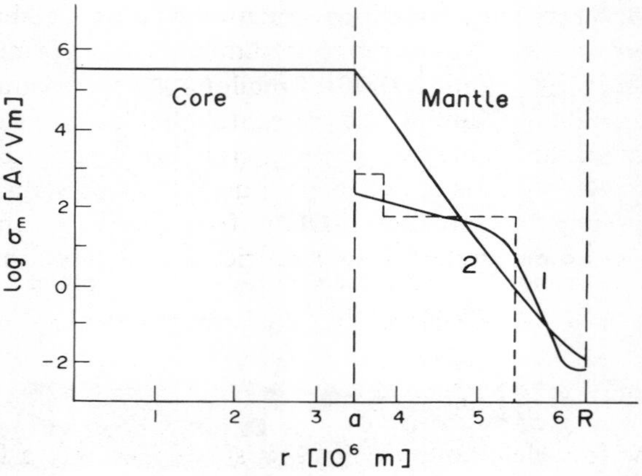

FIGURE 2.3 Electrical conductivity in the Earth's interior. The two solid curves (1) and (2) in the mantle represent results of McDonald and Kolomiytseva, respectively. The *dotted step* function is the conductivity adopted by Roden in his model 9. (Siemens = Ampere/Volt: A/Vm) (according to Stix, 1983).

Recently, Voorhies and Benton (1982) have found, on the basis of MAGSAT data, an even closer value: $r = 3484 \pm 48$ km (see §4.14).

Let us note here that, like all transport properties, electrical conductivity is strongly dependent on temperature and we will see later on (§2.15) that this dependence has the form

$$\sigma = \sigma_0 \exp(-H^*/kT) \tag{2.39}$$

H^* being the activation enthalpy.

The toroidal magnetic field generated in the liquid core (see §4.13) penetrates that part of the lower mantle where the electrical conductivity is sufficiently high and vanishes at the outer boundary of this conducting layer.

To represent the electrical conductivity in the lower mantle, Braginsky and Fishman (1976) and later Stix (1982) have used a modelisation according to the power law:

$$\sigma_r = \sigma_a \left(\frac{a}{r}\right)^\alpha \qquad r > a \text{ (core radius)}$$

They had to use quite high values for α (12 in Braginsky and Fishman; 16 in Backus with $\sigma_a = 3000$ S m^{-1} (1983); 30 in Stix) which results in a concentration of the conductivity in a thin layer close to the core boundary.

2.6 The Grüneisen parameter γ

There are different concepts at the base of its definitions and discussion about the validity of its recent applications in internal geophysics. However this parameter appears in many equations (equation of state, adiabatic gradient, temperature of fusion), and as its range of numerical values is much narrower than all other thermodynamic quantities, it allows simple integrations at least as approximations. As we cannot enter here into all the details of contestations I have included in these notes a recent criticism published in *EOS* in 1981.

I will give two definitions of γ, one from quantum and statistical physical theory (lattice Grüneisen parameter) and one from thermodynamics (thermodynamic Grüneisen parameter). Under restricted conditions they should be equivalent.

The lattice Grüneisen parameter

In a lattice the free Helmholtz energy F is subdivided into a lattice energy (potential energy of a static solid at 0 K) ϕ_0 and a vibrational energy F_{vib} which depends upon specific volume and temperature

$$F(V,T) = \phi_0(V) + F_{\text{vib}}(V,T) \tag{2.40}$$

with the equations of state

$$P(V,T) = -\left(\frac{\partial F}{\partial V}\right)_T \qquad S(V,T) = -\left(\frac{\partial F}{\partial T}\right)_V. \tag{2.41}$$

In a solid, one atom is maintained at its mean position by a combination of attractive (coulombian) forces and repulsive forces. The potential energy, which is the energy of the configuration, is the sum of a negative coulombian term and positive repulsive terms, for example:

$$F(r) = -\frac{A}{r} + \frac{B}{r^n} \tag{2.42}$$

where n depends upon the proposed model (it can be $n = 7$, for example).

The oscillations of an atom inside such a potential well are therefore anharmonic (that is non-sinusoidal): it is easier to spread them than to draw them closer. This, by the way, is the explanation for the thermal dilatation: during their thermal oscillation atoms spend more time at greater separation than closer.

Now in the usual definition of the Grüneisen parameter one takes only the harmonic part of the potential which, according to Einstein (1907), is:

$$F_{\text{vib}}(V,T) = \sum_{i=1}^{3N} \frac{1}{2} h\omega_i + \sum_{i=1}^{3N} kT \ln\left(1 - e^{-h\omega_i/kT}\right) \tag{2.43}$$

k = Boltzmann constant = thermal energy corresponding to 1 Kelvin
ω_i = frequency of vibration
h is for $h/2\pi$ ($h\omega_i$ is the separation of energy levels)
N = the number of particles per mole or Avogadro number.

From this a simple computation gives

$$P(V,T) = -\left(\frac{\partial F}{\partial V}\right)_T = -\frac{d\phi_0}{dV} + \frac{1}{V}\sum_i \gamma_i \varepsilon_i \tag{2.44}$$

where

$$\gamma_i = -\frac{d \ln \omega_i}{d \ln V} = -\frac{V}{\omega_i}\frac{d\omega_i}{dV} \tag{2.45}$$

which is the "*harmonic*" *Grüneisen parameter* (anharmonicity being not considered, see Anderson, 1982) and

$$\varepsilon_i(V,T) = \left(\frac{1}{2} - \frac{1}{1-\exp(h\omega_i/kT)}\right)h\omega_i \tag{2.46}$$

Obviously γ_i are dimensionless numbers.

It is currently admitted that γ is the same for all modes of vibration, that is whatever T (or i).

Mulargia and Boschi (Bologna Istituto di Geofisica) have recently published several papers giving the expression for an anharmonic γ.

The thermodynamic Grüneisen parameter

One has evidently here

$$E_{\text{vib}} = \sum_{i=1}^{3N} \varepsilon_i \tag{2.47}$$

while the internal energy is

$$E = 3NkT \tag{2.48}$$

Thus from (2.44)

$$\left(\frac{\partial P}{\partial T}\right)_V = \frac{\gamma}{V}\left(\frac{\partial E}{\partial T}\right) = \frac{3Nk}{V}\gamma \tag{2.49}$$

because γ as defined here depends only on V and *not of* T (this is the restriction!).

Now the definitions of thermal dilatation (2.26) and incompressibility (2.24) have given:

$$\alpha k_T = \left(\frac{\partial P}{\partial T}\right)_V \tag{2.50}$$

while (2.21) gives:

$$C_V = \left(\frac{\partial E}{\partial T}\right)_V = 3Nk \text{ (Dulong and Petit law)} \tag{2.51}$$

From this we easily derive the *thermodynamic definition of* γ:

$$\boxed{\begin{aligned} \gamma &= \frac{\alpha k_T V}{C_V} = \frac{\alpha k_T}{\rho C_V} \\ &\text{or} \\ \gamma &= \frac{\alpha k_S V}{C_P} = \frac{\alpha k_S}{\rho C_P} \end{aligned}} \tag{2.52}$$

(V being the specific volume, one has $\rho V = 1$)
α, k_T or k_S, C_P or C_S, ρ vary considerably inside the Earth.

On the contrary γ which, according to (2.52) as well as to (2.45) is dimensionless, is found to be between rather narrow limits for all materials (1.2 to 2.6)* and for the conditions inside the liquid core it has been taken as a constant equal to 1.5 (see figure 2.5).

In a recent work O. L. Anderson (1982) shows that in the core and inner core, corrections for anharmonicity and for temperature are not significant (unless $T > 5000$ K which is not the case) and that $\gamma = 1.5$ is a convenient value to adopt. We should also note here that several authors, considering C_V as a constant in some parts of the mantle, are deriving from (2.52) an empirical law

$$\gamma \rho^q = c^{\text{ste}} \tag{2.53}$$

with q slightly higher than unity.

A great number of papers are dealing with the uncertainties related with the use of the Grüneisen parameter. Amongst them the following ones could give an idea of the different arguments in presence: Anderson (1982), Boschi and Mulargia, (1979) Fazio, Mulargia and Boschi (1978), Jamieson and Demarest (1978), Mulargia (1978), Mulargia and Boschi (1979), Stacey (1972).

A short summary of criticisms and objections is given in *EOS*, **62**, 649 (1981).

The Grüneisen parameter as a function of incompressibility and pressure

Vashenko and Zubarev (1963) have demonstrated the formula

$$\gamma = \left(\frac{1}{2}\frac{dk_T}{dp} - \frac{5}{6} + \frac{2}{9}\frac{p}{k_T}\right) \Big/ \left(1 - \frac{4}{3}\frac{p}{k_T}\right) \tag{2.54}$$

on the basis of an analysis of classical three-dimensional atomic oscillations with an anharmonic atomic potential. This equation is in good agreement with

* From laboratory measurements based upon expansion coefficient one has found:

Fe	1.60	Cu	1.96	W	1.62
Co	1.87	Pd	2.23	Pt	2.54
Ni	1.88	Ag	2.40		

laboratory measurements made on materials where the interatomic forces may be considered as central forces.

One may find a rather simple demonstration of this formula by Irvine and Stacey (1975) in the case of a cubic lattice. They consider the small vibrations of two neighbouring atoms when the material is maintained at constant volume by a variable pressure and they develop the mutual forces in Taylor series.

The Vashenko–Zubarev equation is valid only for central forces but this is the case for very compact materials under high pressure, thus for the core but not for the upper mantle.

Irvine and Stacey have introduced a correction term for non-central forces. Stiller and Franck (1980) have shown that under the core conditions the temperature correction to the Vashenko–Zubarev formula is negligible. See also Falzone and Stacey (1981), Irvine and Stacey (1975), Anderson (1982).

Let us remark that for zero pressure one has the very simple relation.

$$\gamma = \frac{1}{2}\frac{dk_T}{dp} - \frac{5}{6} \tag{2.55}$$

which can be verified by laboratory measurements.

Having obtained numerical values of p (eqn. 1.30, figure 1.15), k (eqn. 1.41) and dk/dp (figure 2.4) one can derive γ (figure 2.5) without the need of an equation of state.

2.7 Introduction of the Grüneisen parameter in fundamental relations

The adiabatic gradient of temperature

From eqns (2.31) and (2.52) one immediately obtains

$$\left(\frac{\partial T}{\partial P}\right)_S = +\frac{\alpha VT}{C_P} = \frac{\gamma T}{k_S}$$

or

$$T^{-1}\left(\frac{\partial T}{\partial P}\right)_S = \frac{\gamma}{k_S}$$

so that

$$\frac{d}{dr}(\log T) = -\frac{g\gamma}{\phi} \tag{2.56}$$

which can be integrated as follows:

$$\log\frac{T_2}{T_1} = -\gamma\int_{r_1}^{r_2}\frac{gdr}{\phi} = -\gamma\{I(r_2) - I(r_1)\} \tag{2.57}$$

Thus if we knew the temperature at one of the two boundaries of the liquid core (that is the inner-outer core boundary) we could quite easily calculate the profile of adiabatic temperature inside the core, up to the mantle–core boundary. A very simple evaluation is made with $g \sim 10^{-2}$ km s^{-2}, $T = 2000°$C, $\gamma = 1.5$, $\phi = 60$ km^2 s^{-2}

$$\left(\frac{dT}{dr}\right)_S = -0°5 \text{ km}^{-1} \tag{2.58}$$

Adiabatic and isotherm incompressibilities

Specific heats (2.20), (2.21) are related by:

$$C_P = T\left(\frac{\partial S}{\partial T}\right)_P = T\left(\frac{\partial S}{\partial T}\right)_V + T\left(\frac{\partial S}{\partial V}\right)_T\left(\frac{\partial V}{\partial T}\right)_P$$

with

$$\left(\frac{\partial S}{\partial V}\right)_T = \left(\frac{\partial P}{\partial T}\right)_V = -\left(\frac{\partial V}{\partial T}\right)_P\left(\frac{\partial P}{\partial V}\right)_T$$

thus, by using (2.24) and (2.26):

$$C_P = C_V + T\alpha^2 V k_T$$

while (2.28):

$$\frac{C_V}{k_T} = \frac{C_P}{k_S}$$

and by using (2.52):

$$C_P - C_V = C_V \gamma \alpha T \tag{2.59}$$

$$k_S = k_T(1 + \gamma\alpha T) \tag{2.60}$$

This relation is important because seismological observations give k_S and because the speed of seismic waves is high and does not leave time for heat exchanges, while laboratory measurements and the Vashenko–Zubarev equation are related to k_T.

Thus eqn (2.60) must be associated to the eqn (2.54) for treating the Earth's interior. Obviously iterations will be needed to solve this system and derive γ from seismological observations of k_S, dk_S/dp (see figure 2.4). (Note that $(dk/dP)_S$ is a dimensionless number which is about 3.5 in the lower mantle but less, i.e. 3.0, in the D″ layer: it is 3.6 in the outer core).

2.8 The Clausius Clapeyron equation

From the fundamental thermodynamic relation

$$dE = TdS - PdV + \mu dN \tag{2.61}$$

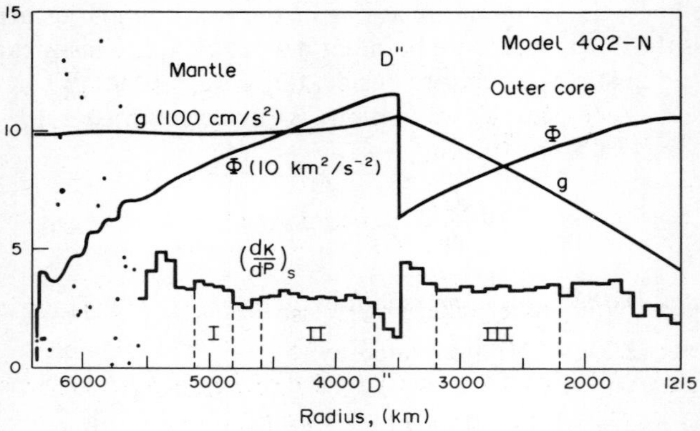

FIGURE 2.4 Variations of dk_S/dp with radius (Butler and Anderson, 1978). Bullen has suggested an empirical "compressibility-pressure relation": $dk_S/dp = 5\text{–}5.6\ p/k_T$ (upper mantle and D″ layer excluded)

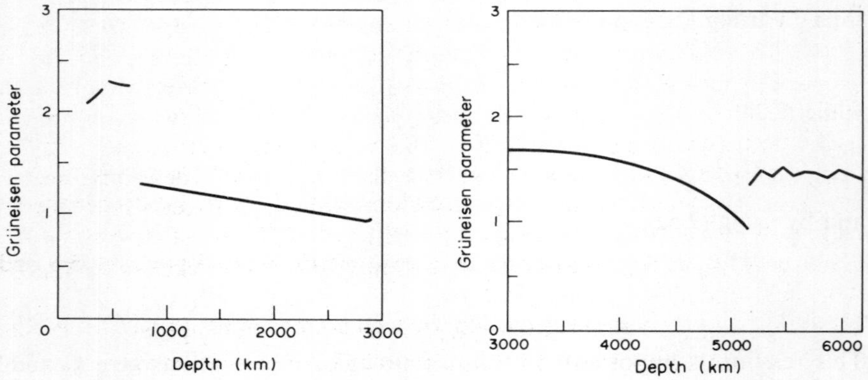

FIGURE 2.5 Thermal Grüneisen paramater γ_{sa} calculated (a) for the mantle below 270 km depth and (b) for the core. Variations such as that at the base of the mantle and in the inner core arise from small variations in differentiation of the Earth model (according to Brown and Shankland, 1981).

we have derived

$$\mu = \left(\frac{\partial E}{\partial N}\right)_{S,V}$$

the chemical potential, the Gibbs free energy being $G = N\mu$. It follows that:

$$N d\mu = -SdT + VdP \tag{2.62}$$

To allow two phases to coexist, their Gibbs free energy must be equal at the melting point, whatever be pressures:

$$\mu_1(P, T) = \mu_2(P, T)$$

thus

$$\left(\frac{\partial \mu_1}{\partial T}\right)_P + \left(\frac{\partial \mu_1}{\partial P}\right)_T \left(\frac{dP}{dT}\right) = \left(\frac{\partial \mu_2}{\partial T}\right)_P + \left(\frac{\partial \mu_2}{\partial P}\right)_T \left(\frac{dP}{dT}\right) \tag{2.63}$$

or, from eqn. 2.62

$$S_2 - S_1 = (V_2 - V_1)\frac{dP}{dT} \tag{2.64}$$

so that the *latent heat* of the transition is

$$L = (S_2 - S_1)T = (V_2 - V_1)T\frac{dP}{dT} \tag{2.65}$$

The latent heat characterizes an exchange of heat at *constant* temperature: melting happens at constant temperature. The liquid at T_m is at a higher level of energy that the solid at same T_m temperature. Thus heat must be furnished to convert a solid to liquid state. Inversely at solidification heat will be released: this is important to understand the formation of the inner core.

The entropy of fusion is

$$\Delta S = S_2 - S_1 = \frac{L}{T_m} \tag{2.66}$$

with

$$\frac{dP}{dT_m} = \frac{L}{T_m \cdot \Delta V}. \tag{2.67}$$

It gives

$$\frac{dT_m}{dP} = \frac{T_m \cdot \Delta V}{L} = \frac{T_m}{L}(\rho_L^{-1} - \rho_S^{-1}) \tag{2.68}$$

this is the *Clausius Clapeyron equation*, which expresses the effect of pressure on the melting temperature T_m, an effect which is of extreme importance for the physics of the Earth's interior. The latent heat of crystallization gives an important contribution provided that the volume decrease is 0.5% or larger.

2.9 Lindemann's law of fusion (1910)

Fusion (or solidification) is an equilibrium state between the liquid phase and the solid phase.

The idea of Lindemann was that melting starts when the mean square root of the amplitude of the thermal atomic vibrations approaches half the distance (or a critical value of it, δ) of the neighbouring atoms. This allows direct collisions and the shear modulus vanishes.

This law is thus obviously based upon solid state physics and therefore is valid on the condition that the liquid preserves some properties of the solid state i.e. the "order" of the structures at small scale; the "disorder" characteristic of a liquid must exist at large scale only. This is acceptable for the structures in the lower mantle and in the core which are close packed because of the pressure.

Let the angular frequency of vibration be $\omega = 2\pi\nu$ and the lattice parameter $a \approx V^{1/3}$. The kinetic energy of the oscillation is to be equal to the Boltzmann energy of melting:

$$E_{\text{kin}} = \tfrac{1}{2}m\delta^2 a^2 \omega^2 = \tfrac{3}{2}kT_f \tag{2.69}$$

Experience shows that δ is more or less constant for all monoatomic solids ($\delta \sim 0.07$).

Let us take the logarithmic derivatives in (2.69):

$$\frac{2}{3} + 2\frac{d \log \omega}{d \log V} = \frac{d \log T_f}{d \log V} \tag{2.70}$$

and remember that the incompressibility coefficient is

$$k_S = -\left(\frac{dP}{d \log V}\right)_S \tag{2.71}$$

while the lattice Grüneisen parameter is

$$\gamma = -\frac{d \log \omega}{d \log V} \tag{2.72}$$

we get

$$\boxed{\frac{1}{T_f}\frac{dT_f}{dP} = \frac{2(\gamma - \frac{1}{3})}{k_S}} \tag{2.73}$$

Stacey and Irvine (1977) have given a solid thermodynamic basis to this law and an atomistic explanation which gives a high level of validity at high pressure that is at the interior of the Earth.

2.10 Chemical constitution of the liquid core

The inversion of seismological data (travel times of the different phases and free oscillation frequencies) made in the presence of some basic constraints of astronomical and geodetical nature (mean radius, mass, moments of inertia) has been made with great success for over 20 years, giving for the liquid core densities systematically lower—by some 10%—than the density of pure liquid iron at the pressure inside the core. Also the bulk modulus of pure iron is larger than that of the core. This obviously indicates the presence of some light

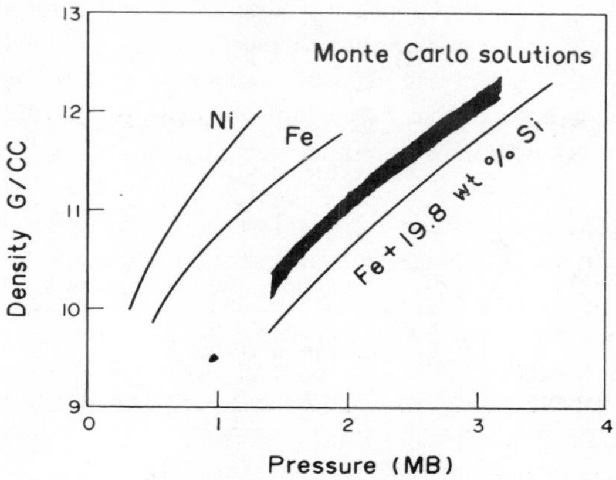

FIGURE 2.6 Band of core densities from successful models together with shock wave density data for Fe, Ni and Fe + 19.8 wt. % Si (Press, 1971).

component associated to iron (with probably a small percentage of nickel). On geochemical grounds it is possible to restrict the number of candidates for this alloy to three; silicon, sulphur, oxygen. After the important work of Usselman (1975), geochemists in general give preference to sulphur but from our point of view it may be sufficient to say that a wide range of chemical compositions may have the same physical properties, the essential one being the phenomenon of fractionated crystallization.

This is best described by considering figure 2.7, which shows how the melting temperature of an alloy Fe + FeS changes in function of the concentration of impurities (here FeS) and has a marked minimum for a proportion of 73 % pure iron with 27 % of sulphur iron.

This minimum point is called the *eutectic point* and the corresponding mixture is called the *eutectic mixture*.

The composition of the Earth's core being close enough to the eutectic composition this explains that one can indeed have a liquid core at a moderate temperature (3800 K). Note that the drawing of figure 2.7 is made for atmospheric pressure.

Pressure has much less effect on the eutectic level than on the melting temperature of the pure elements. Usselman (1975) has shown that when pressure increases the eutectic temperature for Fe, FeS increases and the eutectic composition moves towards a more metallic composition:

$$\text{at 3 Megabar, } T_{eu} = 2000\,^{\circ}\text{C and c}(\%\text{S}) = 15\,\%.$$

To be in agreement with the value of the density and bulk modulus determined by seismology for the liquid core, one must have a mixture slightly more dense, thus more metallic than the eutectic. Consequently if the Earth is cooling the iron-nickel excess to the eutectic composition will precipitate and form the

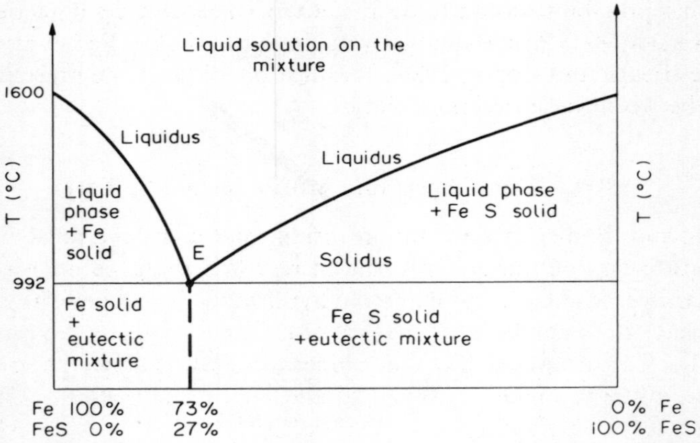

FIGURE 2.7 This representation obeys the Gibbs rule of phases: with n components (here: 2) in r phases (here also 2) the number of thermodynamic degrees of freedom is f with $r+f=n+2$. There are thus 2 degrees of freedom: temperature and concentration are the two variables, while pressure is kept constant.

inner core while the remaining liquid, redistributed by buoyancy in the core, will increase its concentration in sulphur and will very slowly evolve towards the eutectic composition. If this is the process, the inner core will be a pure alloy iron–nickel without light elements: free oscillations data indicate that the inner core is a solid (perhaps not far from melting but solid anyway) with a density higher than the density of the liquid core by about 0.6 g cm^{-3} corresponding to some proportion of iron and nickel (see Table 1.6).

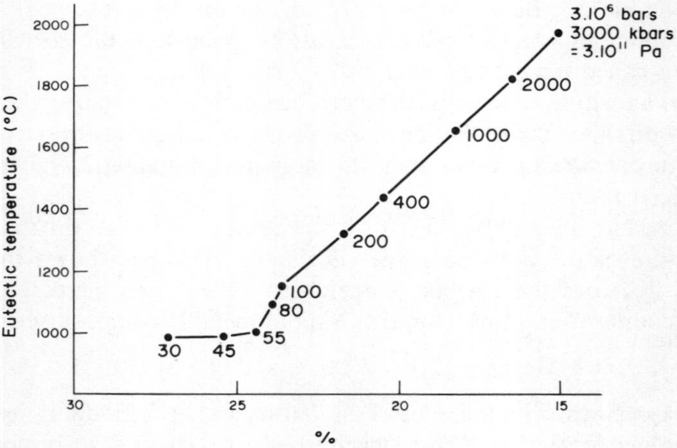

FIGURE 2.8 Eutectic composition and temperature for an Fe–FeS alloy as function of the pressure (Usselman, 1975).

Another possibility could be a Fe–FeO mixture which is eutectic for 25% weight FeO at 90 GPa and qualitatively similar to the Fe–FeS mixture at high pressure (Fearn and Loper, 1983). It is not possible, at the present stage, to choose between these two possibilities.

2.11 The temperature profile inside the Earth

To perform a numerical integration inside the Earth, one must first fix the temperature on one radius. This cannot be other than the inner core–outer core interface, where we can make the hypothesis that the effective temperature is that of the solidus temperature of the quasi-eutectic mixture Fe–FeS.

Another constraint is that the temperature just below the mantle–core interface must be higher or equal to the melting temperature, but this is obviously less constraining.

First of all Stacey (1977b) uses the Lindemann law of melting to extrapolate the laboratory results of Usselman on the alloy Fe–FeS (1975, see figure 2.8) to the pressure 3.2 Megabar existing at the inner core outer boundary.

$$\log(T_{m2}/T_{m1}) = \int_{p_1}^{p_2} \frac{2(\gamma - \frac{1}{3})}{k_S}\, dp \tag{2.74}$$

where

$$2\left(\gamma - \frac{1}{3}\right) = 2.33 \quad \text{if } \gamma = 1.5$$

and, according to eqn (1.38):

$$\int_{p_1}^{p_2} k_S^{-1}\, dp = \log(\rho_2/\rho_1)$$

thus

$$T_{m2}/T_{m1} = (\rho_2/\rho_1)^{2.33} \tag{2.75}$$

which gives

$\underline{T = 4168\,\text{K}}$ at the inner core–outer core interface

Convection acting inside the core is responsible for the permanence of the magnetic field (see §4.13); but convection transports heat outside (or upwards) and in this way reduces a superadiabatic gradient toward the adiabatic value. One can therefore consider that the temperature gradient is practically adiabatic in the core. Then

$$T^{-1}\left(\frac{\partial T}{\partial P}\right)_S = \gamma/k_S \tag{2.76}$$

gives

$\underline{T = 3157\,\text{K}}$ at the core–mantle interface

This corresponds to the gradient 0.5 K km^{-1} as estimated in (2.58).

By again extrapolating Usselman's results one finds about 7 % sulphur at the inner core–outer core boundary and about 10 % sulphur at the mantle–core boundary, this is a general mean of 9 % which explains why the core density is some 10 % less than pure iron density at the core pressure.

In the lower mantle Stacey uses the Vashenko–Zubarev equation to determine γ. However as seismology gives k_S while this equation is based upon k_T one has to use the thermodynamic relation (2.60):

$$k_S = k_T(1 + \gamma\alpha T) \tag{2.77}$$

which already contains γ. Moreover one has to calculate p for the temperature T by using the Mie–Grüneisen equation of state:

$$\left(\frac{\partial P}{\partial T}\right)_V = \alpha k_T = \gamma\rho C_V \tag{2.78}$$

or

$$P_T - P_0 = \int_0^T \gamma\rho C_V dT \sim \gamma\rho E_T \tag{2.79}$$

often called the thermal pressure equation.

It is thus necessary to proceed by successive iterations, and this is simplified because $\partial\gamma/\partial p$ is negligible and because the difference between k_T and k_S is only 5–6 % in the core. Finally Irvine and Stacey (1975) obtain:

$\gamma = 1.4$ with $d\gamma/dp \simeq 0.08$ Mbar^{-1} in the core
$\gamma = 1.0$ in the lower mantle
$\gamma = 0.8$ in the upper mantle, on the basis of laboratory measurements.

Jamieson, Demarest and Schiferl (1978), having introduced electronic and magnetic contributions, conclude that surely

$$1.2 < \gamma < 2.0 \quad \text{in the core.} \tag{2.80}$$

The Stacey evaluation of the solidus temperature at the inner core boundary (4168K) is based upon the Lindemann theory which has theoretical support for pure materials. However the validity of its application to mixtures has not been established while the change of eutectic composition with pressure is not modelled by this theory (Brown and McQueen, 1982).

Moreover the outer core properties should be estimated by a liquid-state theory and not by solid-state theory, but the properties of liquid iron are known only at very low pressures.

By the use of the modern theory of liquids, Stevenson (1980) has derived an equation which is valid for a high-pressure liquid near its melting point:

$$\frac{d \ln T_f}{d \ln \rho} = \frac{\gamma C_V - 1}{C_V - \frac{3}{2}} \tag{2.81}$$

TABLE 2.1 *Temperature profile according to Stacey* (1977)

	r (km)	γ	T_m(K)	T(K)	
Inner core	0	1.186	4.337	4.286	
	1217.1	1.203	4.168	4.168	
	1217.1	1.265	4.168	4.168	
Outer core	1600	1.278	4.043	4.074	
	2000	1.296	3.879	3.947	
	2500	1.325	3.618	3.745	
	3000	1.366	3.283	3.483	
	3485.7	1.419	2.878	3.157	
	3485.7	0.914	3.157	3.157	
Lower mantle	3500	0.919	3.119	3.005	
	4000	0.935	2.986	2.840	
	4500	0.958	2.812	2.662	
	5000	0.985	2.624	2.518	
	5500	1.016	2.415	2.335	
	5701	1.029	2.312	2.250	
	5701	0.736	2.312	2.250	
	5951	0.768	2.110	2.085	
Upper mantle				*Continents*	*Oceans*
	6001	0.776	2.054	2.025	2.036
	6201	0.811	1.775	1.480	1.780
	6360	0.848	1.517	540	550

Several other profiles have, of course, been calculated by different authors. They do not contradict the present one.

where C_V is the atomic contribution to the specific heat in units of Boltzmann's constant, per atom. This equation reduces to Lindemann's law for the classical harmonic oscillator value $C_V = 3$ (see eqns 2.22, 2.73, 2.81) but C_V is, in fact, slightly greater (Anderson, 1982).

On the other hand, an important question would be to determine which crystallographic phase of iron is present inside the inner core. The different phases are:

the normal	α phase	(body-centred cubic), low temperature, low pressure
the	δ phase	(body-centred cubic), high temperature
the	γ phase	(face-centred cubic)
the	ε phase	(hexagonal close-packed), high pressure

and the high-temperature liquid l phase.

Phase transition and discontinuity in elastic wave velocities appear in function of pressure ($\varepsilon \rightarrow \gamma$ at about 200 GPa and $\gamma \rightarrow$ liquid at about 250 GPa). The volume changes associated with these transitions may reduce the discrepancy between liquid iron and core density and consequently affect the interpretations.

Two triple points are well known: (α–γ–ε) at about 115 kilobar, 800K and (δ–γ–l) at 50 kilobar, 2000K, but the localization of the third one (γ–ε–l) is most essential to understand the inner core properties.

The coordinates (P, T) of the (γ–ε–l) triple point are obtained by extrapolation of the γ–ε and γ–l phase boundaries. Divergences between the authors are considerable. Liu (1975) obtains very low values (93.5 GPa, 3239K) and consequently an ε-iron solid inner core, while Anderson (1982) obtains high values (350 GPa, 6000K) and a γ-iron solid inner core.

The density of an ε-iron core is comprised between 13.5 and 14.0 mg m^{-3} while the density of a γ-iron core is comprised between 13.0 and 13.4 mg m^{-3}. This last value is in better agreement with all the models given in Table 1.6, but Masters (1979) considers that the density in the inner core is not better known than to $\pm 5\%$.

The estimation of the melting temperature of iron at the core conditions must take this triple-point position into account.

If the ε-phase is considered as the phase in the inner core, the melting temperature will be some hundred degrees greater than estimated from the α-phase which was considered by most of the authors.

Brown and McQueen (1982) illustrate the situation and its uncertainties with a graph reproduced as figure 2.9, where the variation of temperature at the core–mantle boundary caused by a change in the liquidus at the inner core boundary is shown as a function of sulphur content. This work is based on the high-pressure ε phase of iron and gives, at the core–mantle boundary a temperature of 4000 ± 500K and a composition with 5–10 wt % sulphur.

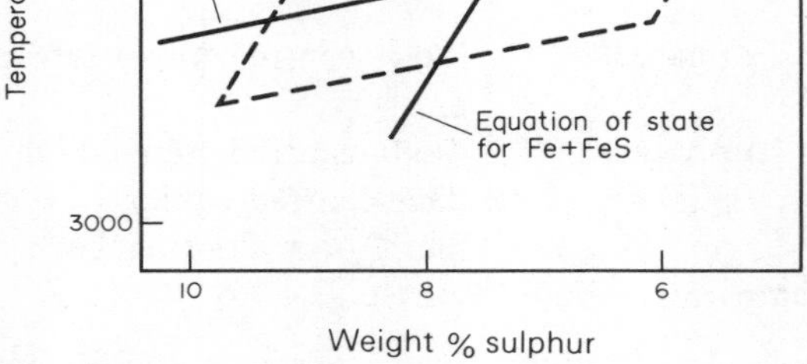

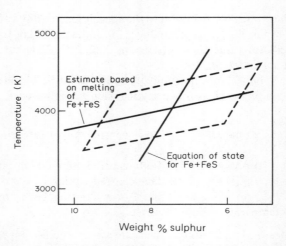

FIGURE 2.9 Melting temperatures estimated for Fe + FeS and temperatures for which given mixtures of Fe + FeS match core densities. The intersection of heavy lines represents a self-consistent solution for temperatures and composition in the core at the core–mantle boundary. Dashed lines indicate uncertainties. (Brown and McQueen, 1982)

This may be compared with a more recent result of Brown, Ahrens and Shampine (1984) who propose 10 ± 4 wt % sulphur and an older one of Ahrens (1979) with 9–12 wt % S.

2.12 Two "core paradoxes"

The Longman paradox (1963)

The system of the six differential equations for the spheroidal deformations of a sphere is now classical, and can be found in many textbooks or papers, but the six parameters involved are given with two different systems of notations:

$H \equiv y_1$: radial displacement	$N \equiv y_2$: radial component of stress
$T \equiv y_3$: tangential displacement	$M \equiv y_4$: tangential component of stress
$R \equiv y_5$: eulerian potential	$L \equiv y_6$: parameter reducing Poisson equation to two differential equations of first order

H, T, R, N, M, L are used by Molodensky and are more mnemonic; therefore we use them here.

If, in the static case (omitting inertial terms $\rho\partial/\partial t^2$), one puts the rigidity μ equal to zero, which seems to be a normal procedure for the Earth's core, one obtains

$$\rho R - \rho g H + \lambda f = 0$$

$$\rho R' + \rho g f - \rho (gH)' + (\lambda f)' = 0$$

(Melchior, *The Tides of the Planet Earth*, p. 99, eqns 5.12 and 5.13). Here the prime corresponds to a derivative along the radius and f is the cubical dilatation factor.

$$f = H' + \frac{2}{r} H - \frac{n(n+1)}{r^2} T$$

Deriving the first and subtracting the second of these equations one obtains

$$f\left(\frac{\lambda \rho'}{\rho^2} - g\right) = 0$$

Thus, either $f = 0$ *which implies an incompressible core*

or $\quad g + \lambda\rho'/\rho^2 = 0$

which is the Adams Williamson equation because $\lambda = k$ when $\mu = 0$!

Thus should the Adams Williamson be automatically obeyed in a static case ($\omega_i = 0$) with $\mu = 0$? Should the core be necessarily an Adams Williamson core? There are no physical reasons for this. The reason for such a paradox is to be ascribed to the fact that it is not permitted to simultaneously put $\mu = 0$,

$\omega_i = 0$ together with a zero viscosity. See, in this respect: Dalhen (1974), Wunsch (1974), Dahlen and Fels (1978).

The Higgins–Kennedy paradox (1971)

There were some years ago many references in the literature to the Higgins–Kennedy "core paradox" which is of another nature. These authors came to a result which showed that the melting point gradient should be smaller than the adiabatic gradient in the Earth's core. This should prevent convection and make impossible the generation of the geomagnetic field which needs a superadiabatic gradient (see fig. 1.26). This view is presently rejected because it would imply, from eqns 2.56 and 2.73, that

$$2\left(\gamma - \frac{1}{3}\right) < \gamma$$

that is
$$\gamma < \frac{2}{3}$$

which cannot be satisfied for the strongly repulsive potentials which are characteristic of the high pressures of the core (eqn. 2.42) (Stevenson, 1980).

2.13 Transport properties—diffusivity and conductivity

To describe the transport of a scalar quantity C, we define its flow $\vec{f}(\vec{x})$ which we can suppose to be simply proportional to the gradient of C. This may be acceptable if this gradient is small enough:

$$f_i = k_{ij} \frac{\partial C}{\partial x_j} \tag{2.81}$$

The transport coefficient appears as a second-order tensor which depends upon the local properties of matter. In an *isotropic* body:

$$k_{ij} = -k\delta_{ij} \tag{2.82}$$

so that one can write

$$\vec{f} = -k\vec{\nabla}C \tag{2.83}$$

which is the equation of diffusion (the flow being directed from high towards low concentrations).

Let us now consider a volume limited by a surface S which contains N molecules. The conservation equation is:

$$\frac{\partial}{\partial t}\iiint_V CN\,dV = \oiint_S \vec{f}\cdot\hat{n}\,ds \tag{2.84}$$

or

$$\iiint_V \left(N\frac{\partial C}{\partial t} - \nabla\cdot k\vec{\nabla}C\right)dV = 0 \tag{2.85}$$

which must be valid for any internal volume, so that if the gradient of k is small ($\nabla \cdot k \nabla C = k \nabla^2 C + \nabla C \cdot \nabla k$) which, in general, is the case:

$$\frac{\partial C}{\partial t} = \chi \nabla^2 C, \quad \chi = \frac{k}{N} \tag{2.86}$$

equation of diffusion for a concentration of salt.

In the case of *heat conduction*, the heat flow is defined by

$$\vec{H} = -k \vec{\nabla} T \tag{2.87}$$

Here, k is the thermal conductivity. Within a short interval of time δt the heat received per unit of mass is

$$\delta Q = -\frac{1}{\rho} \nabla \cdot \vec{H}\, \delta t \tag{2.88}$$

or

$$\delta Q = \frac{1}{\rho} \nabla \cdot (k \nabla T)\, \delta t \tag{2.89}$$

This corresponds to an increase of entropy $\delta S = \delta Q / T$ per unit of mass. At constant pressure we can write (eqn 2.21):

$$T\left(\frac{\partial S}{\partial t}\right)_P = C_P \frac{\partial T}{\partial t} = \frac{1}{\rho} \nabla \cdot (k \nabla T) \tag{2.90}$$

or, finally:

$$\boxed{\frac{\partial T}{\partial t} = \frac{k}{\rho C_p} \nabla^2 T = \chi \nabla^2 T} \tag{2.91}$$

χ is the thermal diffusivity $\{\text{cm}^2\,\text{s}^{-1} = \text{stokes}\}$

k is the thermal conductivity $\{\text{joule}\,\text{s}^{-1}\,\text{cm}^{-1}\,\text{K}^{-1}\}$ or $\{\text{watt}\,\text{cm}^{-1}\,\text{K}^{-1}\}$

C_p is expressed in $\text{cal}\,\text{g}^{-1}\,\text{K}^{-1}$ or $\text{joule}\,\text{g}^{-1}\,\text{K}^{-1}$

In the Earth's core one may guess that

k ~ 0.4 to 0.6 watt $\text{cm}^{-1}\,\text{K}^{-1}$ (Wiedemann–Franz law)

ρ $\sim 11\,\text{g}\,\text{cm}^{-3}$

C_p $\sim 1, 2$ joule $\text{g}^{-1}\,\text{K}^{-1}$

it results that

$$\chi \sim 0.03\text{–}0.05\ \text{cm}^2\,\text{s}^{-1} \tag{2.92}$$

so that, if

$$\nu \sim 0.01\ \text{cm}^2\,\text{s}^{-1}$$

The Prandtl number is:

$$P = \frac{\nu}{\chi} = 0.25 \tag{2.93}$$

2.14 Heat flow from the core

If we accept a value of the electrical conductivity

$$\sigma \cong 5 \times 10^5 \text{ S m}^{-1} \tag{2.94}$$

with a mean temperature

$$T \sim 3500 \text{ K} \tag{2.95}$$

The Wiedemann–Franz law gives us the thermal conductivity

$$\frac{k}{\sigma T} = 2.23 \times 10^{-8} \text{ watts } \Omega \deg^{-2} \tag{2.96}$$

that is

$$k = 39 \text{ watts m}^{-1} \deg^{-1} \tag{2.97}$$

with an adiabatic gradient of about $-0°5 \times 10^{-3} \text{ m}^{-1}$ (eqn (2.58)) one obtains the total heat flow crossing the surface S of the core ($S = 1.53 \times 10^{14} \text{ m}^2$):

$$\vec{H} = -k\frac{dT}{dr} \times S = 3 \times 10^{12} \text{ watts} \tag{2.98}$$

or about 20 milliwatt per square meter.
This is the heat *conducted* down the adiabatic gradient; it represents 10% of the total geothermal flux from the Earth's outer surface (see §2.14) which is not negligible. Thus if the heat flow in the core does not exceed this quantity (which corresponds to a *zero thermodynamic efficiency*) there will be no thermal convection: all heat can be evacuated by simple conduction.

On the other hand the heat flux imposes an upper limit to the conductivity in the core which is surely not more than $\sigma = 10^6 \text{ S m}^{-1}$.

Note also that, if the heat cannot be extracted fast enough from the core, this may lead to the development of a stably stratified region at the top of the outer core within which radial motions would be inhibited.

The volume of the core being about $1.7 \times 10^{20} \text{ m}^3$, the volumic outflow in the core would be 2.10^{-8} watts m^{-3}, which is 100 times less than in acid rocks of the crust but 100 times more than the metallic meteorites.

Some general properties of the Earth's heat flow

The heat flow is defined by the equation

$$H = -k\frac{\partial T}{\partial r} = +k\frac{\partial T}{\partial z} \tag{2.99}$$

(r is the radius vector, z is the depth).

A planetary mean value for the Earth is

$$\langle H \rangle = 1.2 \times 10^{-6} \text{ cal cm}^{-2} \text{ s}^{-1} = 1.2 \text{ hfu} \tag{2.100}$$

(hfu means *heat flow unit*). The transformations

$$\langle H \rangle = (1.2 \times 4.182) \times 10^{-6} \text{ joule s}^{-1} \text{ cm}^{-2} = 50 \text{ erg s}^{-1} \text{ cm}^{-2} \quad (2.101)$$
$$= 50 \times 10^{-7} \text{ watt cm}^{-2} = 50 \text{ milliwatt m}^{-2} = 1 \text{ hfu}$$

allow writing it with the three different units currently used in the literature:

$$1 \text{ hfu} = 10^{-6} \text{ cal cm}^{-2} \text{ s}^{-1} = 41.8 \text{ mW m}^{-2} \quad (2.102)$$

The heat conducted towards the Earth's surface is comprised between 42 and 60 mW m^{-2} with regional anomalies sometimes reaching a factor of 3 or more. Sixty per cent of this heat flow originate in the upper mantle.

Let us put:

$$H(\theta, \lambda) = \sum_{n=0}^{N} \sum_{m=0}^{n} \{A_{nm} \cos m\lambda + B_{nm} \sin m\lambda\} P_n^m(\cos\theta) \quad (2.103)$$

The mean value at degree n is:

$$\left[\sum_{m=0}^{n} (A_{nm}^2 + B_{nm}^2)\right]^{1/2} \quad (2.104)$$

Chapman and Pollack (1975) have developed $H(\theta, \lambda)$ up to the order 18 and found:

$$A_{00} = 59.2 \text{ mW m}^{-2} \quad (2.105)$$

from which we can write

$$\bar{H} = 59.2 \times 10^{-7} \text{ joule s}^{-1} \text{ cm}^{-2} \quad (2.106)$$

that is, for the whole Earth's surface ($S = 5 \times 10^{18} \text{ cm}^2$),

$$H = 3.10^{13} \text{ joule s}^{-1} = 30 \times 10^{12} \text{ watts} \quad (2.107)$$

which corresponds quite well to the distribution given below for the total geothermal flow:

oceanic surface → 11×10^{11} watts
continental surface → 59×10^{11} watts

crust: ~ 7×10^{12} watts
mantle: ~ 16×10^{12}
core: ~ 3×10^{12} (eqn 2.98)

Total: ~ 26×10^{12} watts (2.108)

Thus, for one year (3.15×10^7 s), we obtain:

$$\text{H} = 10^{21} \text{ joule year}^{-1} \quad (2.109)$$

to be compared to the seismic energy released:

$$\varepsilon_{seis} = 4.5 \times 10^{17} \text{ joule year}^{-1} \tag{2.110}$$

The other coefficients of the harmonic development are smaller, particularly the tesseral ones which are less than 1 mW m^{-2}. Zonals and sectorial coefficients are:

$$\begin{array}{lll} A_{10} = -1.66 & & \\ A_{20} = -0.47 & A_{11} = -2.97 & B_{11} = -1.37 \\ A_{30} = 1.42 & A_{22} = -3.03 & B_{22} = 1.40 \\ A_{40} = -2.03 & A_{33} = 2.00 & B_{33} = 1.92 \\ A_{50} = 1.54 & A_{44} = 0.35 & B_{44} = -2.88 \end{array}$$

Amongst many characteristic areas we can indicate some hot zones:

The Red Sea	up to 120 mW m^{-2}
Some areas in Japan	more than 100 mW m^{-2}
Central Massif in France	up to 125 mW m^{-2}
The Vosges	around 100 mW m^{-2}
Tuscany and North Latium	up to 300 mW m^{-2}
Lipari islands	200 to 300 mW m^{-2}

Cold areas, on the contrary, are:

Precambrian terrains	40 mW m^{-2}
Wales (UK)	30 to 35 mW m^{-2}
Emilia Romagna (Italy)	25 mW m^{-2}
South Sicily	35 mW m^{-2}

and a "very cold spot" in Niger, close to Niamey with only 18–22 mW m^{-2}.

As a matter of fact heat flow is strongly correlated with the age of the Earth's crust, being low in the old continental crust (example the Baltic shield, 2500 million year old: about 35 mW m^{-2}) and high in young oceanic crust (up to 150 mW m^{-2}).

One could *define* the lower limit of the lithosphere as the depth where the geotherm cuts the solidus of the mantle. At such a depth one encounters a *low-velocity zone* with high electrical conductivity ($\sigma \sim 0.1$ siemens m^{-1}) and low viscosity ($\nu \sim 10^{21}$ poise). It is commonly called the *asthenosphere*.

It results that a high heat flow indicates a thin lithosphere and a low heat flow a thick lithosphere. We can expect that an empirical relation could give the thickness d of the lithosphere as a function of the measure of heat flow. Such a formula is given by Chapman and Pollack (1977):

$$d = 155\, H^{-1.46} \tag{2.111}$$

or

$$\log H = \log 155 - 1.46 \log d \tag{2.112}$$

and d should also represent the depth of the asthenosphere.

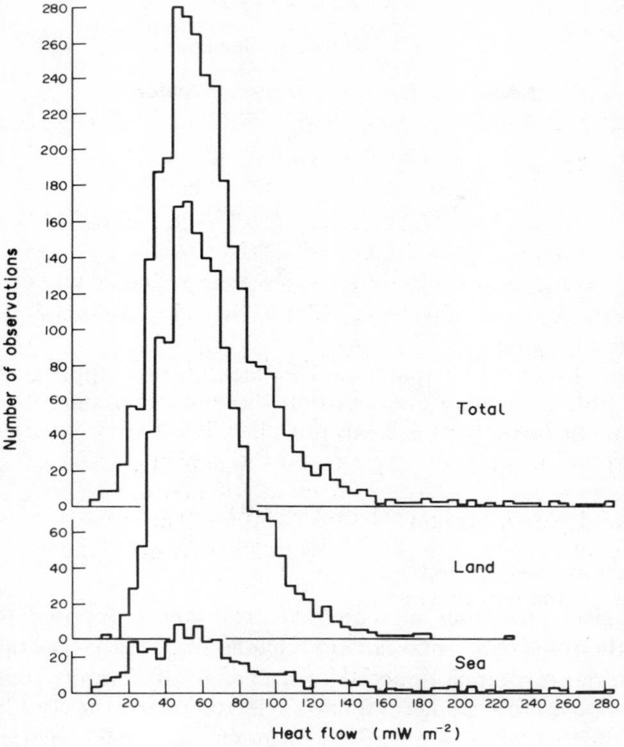

FIGURE 2.10 Histogram of the European heat flow data.

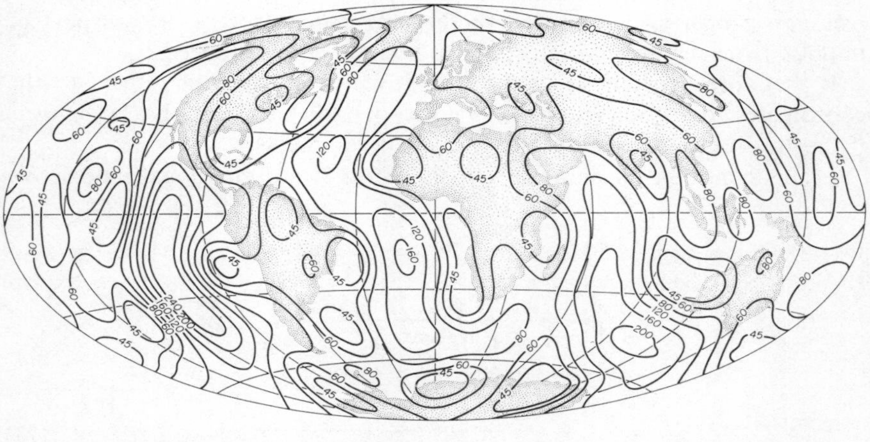

FIGURE 2.11 Degree 12 spherical harmonic representation of global heat flow (after Chapman and Pollack, 1975). Contour values are $mW\,m^{-2}$.

TABLE 2.2

H (mW m^{-2})	Thickness of lithosphere (km) Continents	Oceans
40	250?	150
60	110	85
100	55	55
150	35	35

However continental crust differs from oceanic crust while the solidus depends on the presence of water or volatiles, which considerably decrease the temperature of fusion.

Graphs published by Chapman and Pollack give the approximate figures shown in Table 2.2, where one sees that there is a considerable divergence between the two areas for the weak flow.

2.15 Viscosity

Viscous friction (Navier, 1822)

The rheological properties of matter are resistances opposed to changes: density is the resistance opposed to a change of speed (acceleration), bulk modulus is the resistance opposed to a change of volume, rigidity is the resistance opposed to a change of form (angles). Similarly the viscosity η can be defined as the resistance to a transfer of momentum inside matter.

In laminar conditions transport of momentum occurs by diffusion, on a molecular scale, due to the constant vibration and translation of molecules. The atomistic basis of flow theories rests upon this process. Viscosity is thus a *transport property:* it is related to the *irreversible* process of transport of impulse from regions of high speed to regions of lower speed.

It therefore depends upon the derivatives of speed with respect to the coordinates: for a given gradient of velocity one will have to apply a large stress if the "resistance η" is large. It is a dissipative process.

Let us consider, on figure 2.12 a simplified two-dimensional model. We

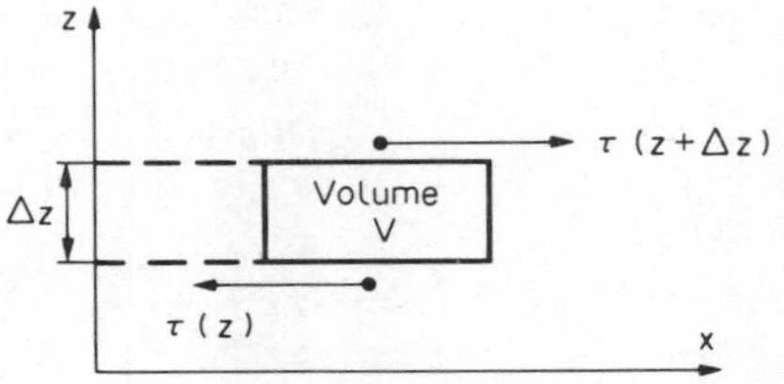

FIGURE 2.12 Viscous friction.

postulate that

$$\tau = \eta \frac{\partial u}{\partial z} \tag{2.113}$$

u being the speed along Ox. The force applied to the volume $V = S\Delta z$ is

$$S\{\tau(z+\Delta z)-\tau(z)\} = S\frac{\partial \tau(z)}{\partial z}\Delta z = V\frac{d\tau}{dz}$$

or, per unit of volume: $\dfrac{d\tau}{dz}$

or, per unit of mass: $\dfrac{1}{\rho}\dfrac{d\tau}{dz} = \dfrac{\eta}{\rho}\dfrac{\partial^2 u}{\partial z^2}$

in the more general form of a three-dimensional flow the viscous stress tensor can be written

$$\bar{\bar{\tau}} = \eta \operatorname{grad} \bar{q} \tag{2.114}$$

and

$$\frac{1}{\rho}\nabla\cdot\bar{\bar{\tau}} = \nu\nabla^2\bar{q} \tag{2.115}$$

here η is called the dynamical viscosity $= \rho\nu$, measured in *poises*, while ν is called the kinematic viscosity, measured in *stokes* ($cm^2\,s^{-1}$).

Viscous flow by diffusion

A model of transport of momentum can be made on the basis of the *activation* theory.

Auto-diffusion is a molecular transport from one region to another region which results in the equalization of concentration (equalization of chemical potentials). It is thus an *irreversible* process and it is a source of *dissipation.* As in the case of thermoconduction the flow is from high concentration (high temperature) towards low concentration (low temperature) regions.

Activation

Diffusion and ionic conductivity processes inside crystals arise from the existence of imperfections and vacancies which can be displaced. However to displace an imperfection from its position of minimum energy and to allow it to jump over the potential barrier which separates it from another minimum energy position one has to furnish an energy which is called the *activated energy* E^*. The process is said to be *activated.* The viscous flow in solids does not result from dislocations but from a migration of atoms across the crystal. Such a migration is thermically activated. Activation energies are about 3 to 5 electron volts ($\sim 7\times10^{-12}$ ergs). The *activation volume* V^* is the most significative property of a vacancy. It differs from the atomic volume because of the relaxation of the atoms surrounding it but it also has a component which results from the movement of the diffusing atoms. It has been measured in the laboratory and is about 10 angström cube for stishovite (SiO_2) or about 11 cm^3/mole.

The diminution of the activation volume with depth decreases viscosity and consequently increases the diffusion.

Thus we have to consider the *activation enthalpy* H^*

$$H^* = E^* + PV^* \tag{2.116}$$

(for the olivine it is 125 ± 5 kcal/mole $\sim$ 6 electron volts).

Temperature and pressure act on the diffusion coefficient according to the Arrhenius law:

$$D = D_0 \exp(-H^*/kT) \tag{2.117}$$

with $D_0 = fa^2\omega$ where f is a number without dimensions which depends from the geometrical structure of the lattice, a is a characteristic length of the lattice, and ω a frequency that one identifies rather arbitrarily to the Debye frequency.

The inversion of this relation allows us to experimentally determine the activation enthalpy which is expressed in electronvolts or kilocalories/mole. Now, for a monoatomic solid Nabarro and Herring have obtained

$$\nu = \frac{kTl^2}{DV_a} \tag{2.118}$$

($l =$ mean radius of grains of the material; $V_a =$ atomic volume) thus

$$\nu = \frac{kTl^2}{D_0V_a} \exp(H^*/kT) \tag{2.119}$$

and the quality factor

$$Q = Q_0 \exp(H^*/kT) \tag{2.120}$$

while the speed of flow is

$$\dot{\varepsilon} = Af(\sigma)\exp(-H^*/kT) \tag{2.121}$$

and the relaxation time

$$\tau = \frac{\nu}{\mu} = \tau_0 \exp(H^*/kT) \tag{2.122}$$

This shows that *for solids*

$$\nu Q^{-1} = \text{constant} \tag{2.123}$$

which empirically is found to be $\sim 4 \times 10^{19}$ (that is $Q = 10^6$, $\nu = 10^{25}$ stokes, $Q = 100$ for $\nu = 10^{21}$ stokes).

Finally, there is of course a relation between diffusion and fusion, and obviously one can empirically show that the activation enthalpy is proportional to the temperature of fusion:

$$H^* = bT_f \tag{2.124}$$

so that

$$D = D_0 \exp(-bT_f(P)/kT) \tag{2.125}$$

$$\nu = \nu_0 \exp(bT_f(P)/kT) \tag{2.126}$$

etc . . .

Experiments by Nachtrieb reconfirm such relations. Stacey has proposed $b \approx 20$.

We can recall here that the electrical conductivity, being also a transport property, will have the form indicated already by eqn (2.39):

$$\sigma = \sigma_0 \exp(-H^*/kT) \tag{2.127}$$

Numerical values

$$\eta = \frac{1}{20}\frac{kTl^2}{Da^3}\, e^{H^*/kT}$$

$l = 10^{-1}$ cm; $a^3 = 10^{-23}$ cm^3; $D_0 = 10$ cm^2 s^{-1}
$k = 1.3806 \times 10^{-23}$ J deg^{-1}
$\eta = 100\, Te^{H^*/kT}$
$H^* = 4$ eV for the upper mantle
$H^* = 6$ eV for the lower mantle

Diffusion and dissipation

All transport properties are controlled by equations of diffusion of the same form:

$$\left.\begin{aligned} \text{momentum:}\quad & \frac{\partial \bar{q}}{\partial t} = \nu \nabla^2 \bar{q} \\ \text{vorticity:}\quad & \frac{\partial \bar{\omega}}{\partial t} = \nu \nabla^2 \bar{\omega} \\ \text{heat:}\quad & \frac{\partial T}{\partial t} = \chi \nabla^2 T \\ \text{magnetic induction:}\quad & \frac{\partial \bar{B}}{\partial t} = \eta^* \nabla^2 \bar{B} \end{aligned}\right\} \tag{2.128}$$

which in, the absence of convection, do not have any source of regeneration.

The coefficients ν, χ, η^* are usually expressed in stokes (cm^2 s^{-1}: surface crossed within 1 second) and their mutual ratios are the *Prandtl dimensionless numbers*:

$$P = \frac{\nu}{\chi} \qquad Pm = \frac{\nu}{\eta^*} \tag{2.129}$$

2.16 Diffusion in a medium composed by two substances: "Onsager relations"

It is important to introduce these relations which obviously govern the situation in the Earth's core. We will simply follow here the Landau and Lifschitz presentation.

We have to introduce a new parameter c, the concentration of substance 1, which is the ratio of its mass to the total mass of the volume we consider. Concentration varies with time for two reasons:

(1) if there is a mechanical macroscopic movement (velocity $\bar{q}$), this is reversible and non-dissipative;
(2) by diffusion which tends to equalize the concentration in all the space; this is *irreversible* and *dissipative*.

If there were no diffusion, the conservation equations would be:

$$\frac{\partial \rho}{\partial t} + \nabla \cdot \rho \bar{q} = 0 \tag{2.130}$$

$$\frac{\partial c}{\partial t} + \bar{q} \nabla c = 0 \tag{2.131}$$

which we may combine (in the case of uniform density) into

$$\frac{\partial}{\partial t}(\rho c) + \nabla \cdot (\rho c \bar{q}) = 0 \tag{2.132}$$

(ρc is obviously the mass of substance 1).

We can also write this

$$\frac{\partial}{\partial t} \int_V \rho c \, dV = -\oint_S \rho c \bar{q} \, dS \tag{2.133}$$

where we should now add a diffusion flow $\vec{i}$ across S:

$$\frac{\partial}{\partial t} \int_V \rho c \, dV = -\oint_S \rho c \bar{q} \, dS - \oint_S \vec{i} \, dS \tag{2.134}$$

or

$$\frac{\partial}{\partial t}(\rho c) = -\nabla \cdot (\rho c \bar{q}) - \nabla \cdot \vec{i} \tag{2.135}$$

which becomes by using (2.130):

$$\rho \frac{\partial c}{\partial t} + \rho \bar{q} \cdot \nabla c + \nabla \cdot \vec{i} = 0 \tag{2.136}$$

For a mixture of the two substances of respective masses m_1 and m_2:

$$c = N_1 m_1 \quad N_1 m_1 + N_2 m_2 = 1 \tag{2.137}$$

Thus, from (2.12):

$$dE = TdS - PdV + \mu_1 dN_1 + \mu_2 dN_2 = TdS - PdV + \left(\frac{\mu_1}{m_1} - \frac{\mu_2}{m_2}\right) dc \qquad (2.138)$$

and, by putting

$$\mu = \frac{\mu_1}{m_1} - \frac{\mu_2}{m_2} \qquad (2.139)$$

while

$$\rho V = 1 \quad PdV = -P\, d\rho/\rho^2$$

we get for the internal energy

$$dE = TdS + Pd\rho/\rho^2 + \mu dc \qquad (2.140)$$

and for the enthalpy

$$dH = TdS + dP/\rho + \mu dc \qquad (2.141)$$

from which we derive

$$dP = \rho dH - \rho TdS - \rho\mu dc \qquad (2.142)$$

so that we have to include a term $\rho\mu dc$ or, according to (2.136):

$$\rho\mu\left(\frac{\partial c}{\partial t} + \bar{q}\cdot\nabla c\right) = -\mu\nabla\cdot\vec{i} \qquad (2.143)$$

in the thermoconduction equation where a term

$$\nabla\cdot\bar{Q} - \mu\nabla\cdot\bar{i} = \nabla\cdot(\bar{Q} - \mu\bar{i}) + \bar{i}\nabla\mu \qquad (2.144)$$

now appears. Q is the heat flow.

Then *the variation of entropy* is:

$$\rho T\left(\frac{\partial S}{\partial t} + \bar{q}\cdot\nabla S\right) = \nabla\cdot\bar{q}\bar{\bar{\tau}} - \nabla\cdot(\bar{Q} - \mu\bar{i}) - \bar{i}\nabla\mu \qquad (2.145)$$

which, combined with Navier and continuity equations will form the system of equations of the problem.

To solve such equations one should finally define the flows of substance and of heat in function of the gradients of concentration and temperature.

If these gradients are small one can admit linear relations *with negative coefficients* because the flow always goes from high values (T or c) to low values.

Then:

$$\bar{i} = -\alpha\nabla\mu - \beta\nabla T \qquad (2.146)$$

$$\bar{Q} - \mu\bar{i} = \beta T\nabla\mu - \gamma\nabla T \qquad (2.147)$$

$$(\alpha > 0, \quad \beta > 0, \quad \gamma > 0)$$

These are the *Onsager relations*. One will find in Landau and Lifschitz (§58) the demonstration that the same coefficient β appears as indicated, and that the flows do not depend upon pressure.

One will observe that in a case where the relation between concentration and temperature is such that:

$$\alpha \nabla \mu + \beta \nabla T = 0 \tag{2.148}$$

there is *no diffusion* at all. One gradient is stabilizing (for example a subadiabatic gradient of temperature, eqn 1.54) and inhibits the effect of the other which is destabilizing.

The elimination of $\nabla \mu$ between the Onsager relations gives

$$\bar{Q} - \left(\mu + \frac{\beta T}{\alpha} \right) \bar{i} = -k \nabla T \tag{2.149}$$

with

$$k = \gamma - \frac{\beta^2 T}{\alpha} \tag{2.150}$$

which is the thermal conductivity.

Status of the Grüneisen Constant

[From *EOS*, (Transactions of the American Geophysical Union), **62,** 35, 649 (1981).]

As a special workshop at the recent meeting of the International Association of Seismology and Physics of the Earth's Interior (IASPEI) in London, Ontario, geophysicist H. H. Schloessin organized a sequence of discussions on 'gamma,' the thermodynamic parameter first derived in 1926 by E. Grüneisen. The Grüneisen parameter is known to all geophysicists as the less than perfect constant that relates volume and thermal energy properties of a substance in geophysical systems at high pressure, ranging from shock-wave environments to that along the temperature gradient of the Earth's lower mantle and core. Derivations and applications of Grüneisen's gamma lean heavily on the fundamentals of quantum and statistical physical theory. The necessary simplicity of the atomic systems for which the scope of the theory applies, and the incredible complexity of its very theoretical basis, have severely limited gamma's practicality for most substances. Furthermore, the geophysical literature is always abreast of the latest cases where a given application of gamma has again failed.

Why, then, does a concept of such apparent tenuity persist? The answer could be made apparent only after considerable esoteric discussion by gamma experts at the IASPEI workshop. New reasons to think gamma is less than constant at high temperature and pressure were provided as usual. Also, as usual, it was clear that gamma survives again in geophysics, status quo.

Thermodynamically, the Grüneisen parameter is a first-law relationship. Statistically, if the Debye theory assumption that each of the vibrational frequencies of the normal modes of crystals vary in proportion to the inverse gamma (γ) power of the volume, then $\gamma = (d \ln \nu^{max} / d \ln V)$ where ν^{max} is the limiting value of the frequency spectrum of a solid, and V is the specific volume. Translating this to thermal-volume terms, $\gamma = (\alpha V / \beta C_v)$ where α is thermal expansion. $(1/V)(\partial V/\partial T)_P$, β is the compressibility $-(1/V)(\partial V/\partial P)_T$, and C_V is the heat capacity at constant volume. It turns out that many assumptions of both theories are unjustified (for one, contributions of the electronic frequencies of a crystal must be considered as well as thermal frequencies), and the interpretations become complex.

The reason that gamma continues to be so hardy and resistant to heavy criticism is that it appears in many equations of state (i.e. the Mie-Grüneisen equation): it appears in the most valuable equations of state for materials at high temperatures and pressures. Whether gamma is a universal constant or not under the entire range of temperatures and pressures within planetary interiors is not important so long as it can be evaluated within certain structural regimes. Two independent approaches to the theoretical-empirical evaluation of the Grüneisen parameter were described in separate discussions at the workshop by O. Anderson and D. Stevenson. A good portion of the discussion was devoted to analysis of where the approximations were not justified and to equations of state with and without gamma. Nonetheless, the discussions left the concept of gamma better understood and relatively unscathed.

It was appreciated at the workshop that there are several considerations about gamma, whether it is to be evaluated for upper- or lower-mantle conditions or for the liquid or solid core. Aside from deducing gamma from theory and from geophysical data, new methods of measuring gamma of materials in the laboratory were discussed. The quest for gamma of the deep earth's interior continues.

Bibliography

Achade, J., Courtillot, V., Ducruix, J. and Le Mouel, J. L. (1980) The late 1960's secular variation impulse: further constraints on deep mantle conductivity. *Phys. Earth Plan. Int.*, **23,** 72–5.

Adler, B. J. and Trigueros, M. (1977) Suggestion of a eutectic region between the liquid and solid core of the Earth. *J. Geophys. Res.*, **82,** 2535–9.

Ahrens, T. J. (1979) Equations of state of iron sulfide and constraints on the sulfur content of the Earth. *J. Geophys. Res.*, **84,** 985–98.

Anderson, D. L. and Hart, R. S. (1976) An Earth model based on free oscillations and body waves. *J. Geophys. Res.*, **81,** 1461–75.

Anderson, O. L. (1982) Are anharmonicity corrections needed for temperature-profile calculations of interiors of terrestrial planets. *Phys. Earth Plan. Int.*, **29,** 91–104.

Anderson, O. L. (1982) The Earth's core and the phase diagram of iron. *Phil. Trans. R. Soc. Lond.* **A306,** 21–35.

Anderson, O. L. (1986) Properties of iron at the Earth's core conditions. *Geophys. J. R. Astr. Soc.*, **84,** 561–79.

Backus, G. (1983) Application of mantle filter theory to the magnetic jerk of 1969. *Geophys. J. R. Astr. Soc.*, **74,** 713–46.

Barton, M. A. and Stacey, F. D. (1985) The Gruneisen parameter at high pressure – A molecular dynamical study. *Phys. Earth Plan. Int.*, **39,** 167–77.

Boschi, E. and Mulargia, E. (1979) The dependence of the melting temperatures of iron upon the choice of the interatomic potential. *Geophys. J. R. Astr. Soc.*, **58,** 204–8.

Braginsky, S. I. and Fishman, V. M. (1976) Electromagnetic interaction of the core and mantle when electrical conductivity is concentrated near the core boundary. *Geomag. Aeron.*, **16,** 443–6.

Brennan, B. J. (1979) A thermodynamically based equation of state. *J. Geophys. Res.*, **84,** 5535–9.

Brown, J. M., Ahrens, T. J. and Shampine, D. L. (1984) Hugoniot data for Pyrrhotite and the Earth's core. *J. Geophys. Res.*, **89,** 6041–8.

Brown, J. M. and McQueen, R. G. (1982) The equation of state for iron and the Earth's core. *Adv. Earth Plan. Sci.*, **12,** 611–23.

Brown, J. M. and Shankland, T. J. (1981) Thermodynamic parameters in the Earth as determined from seismic profiles. *Geophys. J. R. Astr. Soc.*, **66,** 579–96.

Butler, R. and Anderson, D. L. (1978) Equation of state fits to the lower mantle and outer core. *Phys. Earth Plan. Int.*, **17,** 147–62.

Chapman, D. S. and Pollack, H. N. (1975) Global heat flow: a new look. *Earth Plan. Sci. Lett.*, **28,** 23–32.

Chapman, S. and Pollack, H. N. (1977) Regional geotherms and lithospheric thickness. *Geology*, **5,** 265–8.

Falzone, A. J. and Stacey, F. D. (1981) Second order elasticity theory: an improved formulation of the Gruneisen parameter at high pressure. *Phys. Earth Plan. Int.*, **24,** 284–90.

Fazio, D., Mulargia, F. and Boschi, E. (1978) Thermodynamical Gruneisen's function for pure iron at the Earth's core conditions. *Geophys. J. R. Astr. Soc.*, **53,** 531–9.

Fearn, D. R. and Loper, D. E. (1981) Compositional convection and stratification of Earth's core. *Nature*, **289,** 393–4.

Fearn, D. R. and Loper, D. E. (1982) Thermal constraints on the growth of a eutectic inner core. *Geophys. Fluid. Dyn. Inst.*, Florida State University.

Fearn, D. R. and Loper, D. E. (1985) Pressure freezing of the Earth's inner core. *Phys. Earth Plan. Int.*, **39,** 5–13.

Gardiner, R. B. and Stacey, F. D. (1971) Electrical resistivity of the core. *Phys. Earth Plan. Int.*, **4,** 406–10.

Garland, G. D. (1981) The significance of terrestrial electrical conductivity variations. *Ann. Rev. Earth Plan. Sci.*, **9,** 147–74.

Gough, D. I. (1973) The interpretation of magnetometer array studies. *Geophys. J. R. Astr. Soc.*, **35,** 83–98.

Gubbins, D., Masters, T. G. and Jacobs, J. A. (1979) Thermal evolution of the Earth's core. *Geophys. J. R. Astr. Soc.*, **59,** 57–100.

Hide, R. and Malin, S. R. C. (1981) On the determination of the size of the Earth's core from observations of the geomagnetic secular variation. *Proc. R. Soc. Lond.*, **A374,** 15–33.

Higgins, G. and Kennedy, G. C. (1971) The adiabatic gradient and the melting point gradient in the core of the Earth. *J. Geophys. Res.*, **76**(8), 1870–7.

Irvine, R. D. and Stacey, F. D. (1975) Pressure dependence of the thermal Gruneisen parameter, with the application to the Earth's lower mantle and outer core. *Phys. Earth Plan. Int.*, **11,** 157–65.

Jamieson, J. C., Demarest, H. H. and Schiferl, D. (1978) A reevaluation of the Grueneisen parameter for the Earth's core. *J. Geophys. Res.*, **83**(B12), 5929–36.

Johnston, M. J. S. and Strens, R. G. J. (1973) Electrical conductivity of molten Fe-Ni-S-C- core mix. *Phys. Earth Plan. Int.*, **7,** 217–18.

Landau, L. and Lifschitz, E. (1967) *Statistical Physics*. MIR, Moscow.

Lindemann. F. A. (1910) Uber die Berechnung Molekularer Eigefrequenzen. *Physik. Zeit.*, **14** (Jahrg 11), 609–612.

Liu Lin-Gun (1975) On the gamma–epsilon–l triple point of iron in the Earth's core. *Geophys. J. R. Astr. Soc.*, **43,** 697–705.

Loper, D. E. (1984) Structure of the core and lower mantle. *Adv. Geophys.* **26,** 1–27.

Loper, D. E. and Roberts, P. H. (1980) On the motion of an iron-alloy core containing a slurry. *Geophys. Astr. Fluid Dyn.*, I: **9,** 289–321; II: **16,** 83–127.

Loper, D. E. and Roberts, P. H. (1981) A study of conditions at the inner core boundary of the earth. *Phys. Earth Plan. Int.*, **24,** 302–7.
Masters, G. (1979) Observational constraints on the chemical and thermal structure of the Earth's deep interior. *Geophys. J. R. Astr. Soc.*, **57,** 507–34.
Melchior, P. (1982) Physique et dynamique planétaires. *Phys. de L'Int. de la Terre-BXL*, Vol. 5 (Fasc. 1).
Mulargia, F. (1978) On the theoretical evaluation of the Gruneisen function in the harmonic approximation. *J. Geophys. Res.*, **83,** 1843–51.
Mulargia, F. and Boschi, E. (1979) Equations of state of close-packed phases of iron and their implications for the Earth's core. *Phys. Earth Plan. Int.*, **18,** 13–18.
Mulargia, F. and Boschi E. (1980) The problem of the equation of state in the earth's interior. *Phys. Earth's Int.* (Soc. Italiana di Fis.), 337–61.
Mulargia, E., Broccio, F. and Dragoni, M. (1984) On the temperature dependence of the Gruneisen function. *Boll. Geof. Teorica appl.*, **XXVI**(104), 229–36.
Poirier, J. P. and Liebermann, R. C. (1984) On the activation volume for creep and its variation with depth in the Earth's lower mantle. *Phys. Earth Plan. Int.*, **35,** 283–93.
Pollack, H. N. and Chapman, D. S. (1977) Mantle heat flow. *Earth Plan. Sci. Lett.*, **34,** 174–84.
Press, F. (1971) An introduction to earth structure and seismotectonics *Proc. int. school physics.* E. Fermi—Mantle and core in Planetary Physics. Academic Press, 209–41.
Roberts, R. G. (1983) Electromagnetic evidence for lateral inhomogeneities within the Earth's upper mantle. *Phys. Earth Plan. Int.*, **33,** 198–212.
Schmit, U. (1985) Latent-heat driven convection in the Earth's core numerical results. *J. Geodynam.*, **2,** 23–34.
Smylie, D. E. (1965) Magnetic diffusion in a spherically-symmetric conducting mantle. *Geophys. J. R. Astr. Soc.*, **9,** 169–84.
Stacey, F. D. (1972) Properties of the Earth's core. *Geophys. Surv.*, **1,** 99–119.
Stacey, F. D. (1977a) Applications of thermodynamics to fundamental earth physics. *Geophys. Surv.* **3,** 175–204.
Stacey, F. D. (1977b) A thermal model of the earth. *Phys. Earth Plan. Int.*, **15,** 341–8.
Stacey, F. D. and Irvine, R. D. (1977) Theory of melting: thermodynamic basis of Lindemann's law. *Austr. J. Phys.*, **30,** 631–40.
Stevenson, D. J. (1980) Applications of liquid state physics to the Earth core. *Phys. Earth Plan. Int.*, **22,** 42–52.
Stiller, H. and Franck, S. (1980) A generalization of the Vashenko-Zubarev formula for the Gruneisen parameter. *Phys. Earth. Plan. Int.*, **22,** 184–8.
Stix, M. (1982) On electromagnetic core–mantle coupling. *Geophys. Astr. Fluid Dyn.* **21,** 303–13.
Stix, M. (1983) The rotation and the magnetic field of the Earth. *Tidal friction and the Earth's rotation.* II. Springer-Verlag, Berlin and Heidelberg, 89–116.
Usselman, T. M. (1975) Experimental approach to the state of the core. Mantle and outer core. *Am. J. Sci.*, **275,** 291–303.
Vashenko, V. Y. and Zubarev, V. N. (1963) Concerning the Gruneisen constant. *Sov. Solid State*, **5,** 653–5.
Verhoogen, J. (1961) Heat balance of the Earth's core. *Geophys. J. R. Astr. Soc.*, **4,** 276–81.
Verhoogen, J. (1973) Thermal regime of the Earth's core. *Phys. Earth Plan. Int.*, **7,** 47–58.
Voorhies, C. V. and Benton, E. R. (1982) Pole-strength of the Earth from MAGSAT and magnetic determination of the core radius. *Geophys. Res. Lett.*, **9,** 258–61.
Wolf, G. H. and Jeanloz, R. (1984) Lindemann melting law—anharmonic correction and test of its validity for minerals. *J. Geophys. Res.* **89,** 7821–35.
Zharkov, V. N. (1985) Thermal state and thermal regime of the Earth's interior. *Phys. Earth Plan. Int.* **41,** 133–7.
Zharkov, V. N., Karpov, P. B. and Leontjev, V. V. (1985) On the thermal regime of the boundary layer at the bottom of the mantle. *Phys. Earth Plan. Int.* **41,** 138–42.

CHAPTER 3

Hydrodynamics

3.1 Lagrangian and Eulerian descriptions

When, from your living-room, you are looking to the snow falling in slow large flakes, you can do it in two ways:

1. You fix a small area of your window and see the different flakes crossing your sight field at different times, one after the other: this is a *Eulerian* description of the movement and you have simply to express the derivation with respect to time $\partial/\partial t$ at this fixed point of space (the coordinates of the point are independent of time).
2. You keep your eyes upon one well defined flake (starting with an initial position) and you follow its movement across the broad field of your window: this is a *lagrangian* description, the coordinates depend on time, and you have to operate a total derivative with respect to time for a mobile well defined element:

$$\frac{d}{dt} = \frac{\partial}{\partial t} + \frac{\partial}{\partial x}\frac{dx}{dt} + \frac{\partial}{\partial y}\frac{dy}{dt} + \frac{\partial}{\partial z}\frac{dz}{dt}$$

or
$$\frac{d}{dt} = \frac{\partial}{\partial t} + \bar{q}\cdot\overline{\text{grad}} \tag{3.1}$$

$$\bar{q}\,(u, v, w) \text{ is the velocity: } u = \frac{dx}{dt},\ v = \frac{dy}{dt},\ w = \frac{dz}{dt}$$

The lagrangian description is a natural one for elastic solids because the initial position is normally known. It is not suitable for fluids.

The eulerian description is convenient for fluids.

The eqn (3.1) allows to exchange the lagrangian into the eulerian description. d/dt is called material derivative; $\bar{q}\cdot\overline{\text{grad}}$ is the advection operator. For any property X in the velocity field, we can write a Taylor development:

$$X(\vec{P},\tau) = X(\vec{M},t) + \left[\frac{\partial X}{\partial t} + \sum_i \frac{\partial X}{\partial x_i}\cdot\frac{dx_i}{dt}\right]\cdot(\tau - t) + \ldots\ldots$$

from which

$$\lim_{\tau = t} \frac{X(\vec{P},\tau) - X(\vec{M},t)}{(\tau - t)} = \frac{dX}{dt} = \frac{\partial X}{\partial t} + \sum_i u_i \frac{\partial X}{\partial x_i}$$

Thus

$$\frac{dq}{dt} = \frac{\partial q}{\partial t} + (\bar{q}\cdot \text{grad})\bar{q}$$

$$\frac{dT}{dt} = \frac{\partial T}{\partial t} + (\bar{q}\cdot \text{grad})T \tag{3.2}$$

etc . . .

Euler equation

One obtains the equation of motion by equating the force per unit volume $\bar{F} = -\text{grad } P$ to the product of the mass of this unit of volume by its acceleration $d\bar{q}/dt$:

$$\rho \frac{d\bar{q}}{dt} = -\text{grad } P \tag{3.3}$$

which will be written

$$\frac{\partial \bar{q}}{\partial t} + (\bar{q}\cdot\nabla)\bar{q} = -\frac{1}{\rho}\nabla P + \text{other eventual forces} \tag{3.4}$$

in the eulerian description. This is the Euler equation.

Equations for a rotating body

In an inertial frame of reference we have

$$\bar{q}_{\text{in}} = \bar{q}_{\text{rot}} + \bar{\Omega}\Lambda\bar{r} \tag{3.5}$$

$\bar{r}(x, y, z)$ being the vector position,

$$(\partial/\partial t)_{\text{in}} = (\partial/\partial t)_{\text{rot}} + \bar{\Omega}\Lambda \tag{3.6}$$

and the Euler's equation becomes:

$$\rho[\partial\bar{q}_{\text{in}}/\partial t + (\bar{q}_{\text{in}}\cdot\nabla)\bar{q}_{\text{in}}] = -\nabla P - \rho\nabla\phi \tag{3.7}$$

ϕ being the potential of external forces.
Let us substitute $\bar{q}_{\text{in}}$ by (3.5) and put now $\bar{q}_{\text{rot}} \equiv \bar{q}$:

$$\frac{\partial \bar{q}}{\partial t} + (\bar{q}\cdot\nabla)\bar{q} + 2\bar{\Omega}\Lambda\bar{q} = -\nabla p \tag{3.8}$$

where

$$p = \frac{P}{\rho} + \phi - \frac{1}{2}|\bar{\Omega} \wedge \bar{r}|^2 \tag{3.9}$$

is called the *reduced pressure*.
$2\bar{\Omega} \wedge \bar{q}$ is the Coriolis acceleration, $\frac{1}{2}\rho|\bar{\Omega} \wedge \bar{r}|^2$ is the centrifugal force potential.

3.2 Equations of conservation

These equations describe the balance of extensive variables (mass, momentum, energy) inside a given volume. They are written in function of the conjugate intensive variables (density, velocity, temperature, pressure) which are the unknowns in the problem.

The objective is to determine these intensive variables as functions of position and time (x_j, t) in the fluid domain. To achieve this goal one has, however, to introduce phenomenological equations: either experimental laws of behaviour or laws deduced from statistical mechanics.

A conservation equation ascribes the accumulation of an extensive variable per unit of time inside a volume V to a flux of this variable across the boundary of V plus the *production* of this variable inside V. This is why we need a phenomenological equation describing the production of the variable concerned.

Conservation of mass: continuity equation

The mass conservation is expressed as follows

$$\frac{\partial}{\partial t}\iiint_V \rho\, dV + \oiint_S \rho(\bar{q}\cdot\hat{n})dS = 0 \tag{3.10}$$

$\rho\bar{q} = \bar{j}$ is the density of flux of mass.
Applying the Ostrogradsky theorem to (3.10) gives

$$\frac{\partial}{\partial t}\iiint_V \rho\, dV + \iiint_V \operatorname{div}(\rho\bar{q})\, dV = 0 \tag{3.11}$$

which is to be valid whatever be the volume V under consideration. Therefore:

$$\left.\begin{aligned} &\frac{\partial\rho}{\partial t} + \operatorname{div}(\rho\bar{q}) = 0 \\ \text{that is:}\quad &\frac{\partial\rho}{\partial t} + \rho\operatorname{div}\bar{q} + (\bar{q}\cdot\operatorname{grad})\rho = 0 \\ \text{or:}\quad &\frac{d\rho}{dt} + \rho\operatorname{div}\bar{q} = 0 \end{aligned}\right\} \tag{3.12}$$

by introduction of the material derivative.

We see here clearly that the "divergence" is the convergence of flow towards the interior of the volume with its sign changed. The equation of state is expressed by the pressure change:

$$\frac{\partial p}{\partial t} + \bar{q}\cdot\nabla p = -k\nabla\cdot\bar{q} \tag{3.13}$$

with

$$\left.\begin{aligned} \nabla p &= \rho\bar{g} \\ \bar{g} &= -g\hat{r} = -\nabla W \end{aligned}\right\} \tag{3.14}$$

Incompressibility, solenoidal velocity field

For an incompressible fluid ($k = \infty$)

$$\frac{d\rho}{dt} = 0, \quad \text{thus by eqn (3.12):} \quad \text{div } \bar{q} = 0 \tag{3.15}$$

which keeps the pressure finite along the motion (eqn 3.13)

The *incompressibility condition* can be written:

$$\frac{\partial u}{\partial x} + \frac{\partial v}{\partial y} + \frac{\partial w}{\partial z} = 0 \tag{3.16}$$

in cylindrical coordinates:

$$\frac{1}{r}\frac{\partial}{\partial r}(ru) + \frac{\partial v}{r\partial\lambda} + \frac{\partial w}{\partial z} = 0 \tag{3.17}$$

in spherical coordinates:

$$\frac{1}{\sin\theta}\frac{\partial}{\partial\theta}(\sin\theta\, v) + \frac{\partial w}{\sin\theta\partial\lambda} + \frac{1}{r}\frac{\partial}{\partial r}(r^2 u) = 0 \tag{3.18}$$

u, v, w being the components of the velocity along increasing r, θ, λ. Incompressibility excludes the existence of sound waves. It is thus only for velocities which are much lower than the speed of sound (compressibility longitudinal waves), which is the case for the westward drift of the geomagnetic field, that one usually accepts the incompressibility condition which defines a *solenoidal* velocity field:

$$\bar{q} = \text{curl curl } (\bar{r}\phi) + \text{curl } (\bar{r}\psi) \tag{3.19}$$

ϕ and ψ being two scalar functions of the coordinates. This velocity field contains a toroidal part (ψ) and a poloidal part (ϕ).

However the hypothesis of incompressibility, which is universally made in core magnetohydrodynamics, is subject to controversy. Smylie and Rochester (1981) object to the validity of the assumption of a solenoidal flow for motions of large radial length scale because the effects of the pre-existing pressure gradient on the change of density cannot be neglected. We have seen indeed that, at the core–mantle boundary, the pressure is 1.37 megabar and the density 9.9 g cm^{-3}, while at the inner–outer core boundary they are respectively 3.27 megabar and 12.2 g cm^{-3} (see Tables 1.2 and 1.6).

Thus, for motions of large scale, effects of compression and dilatation cannot be neglected: errors in assuming solenoidal flow can be large.

In addition to the effects of compression, self-gravitation changes which accompany motions in a non-neutrally stratified fluid as thick as the Earth's liquid core must be taken into account by adding the Poisson equation into the system of governing equations. This approximation is called the *subseismic approximation* by Smylie and Rochester (1981) (See also Todoeschuck and Rochester, 1980; Hide, 1981; Friedlander, 1985).

The incompressibility assumption raises another problem with respect to the definition of the Brunt Väisälä frequency (eqn 1.45) which, when $k = \infty$, becomes:

$$N^2 = -g(r)\rho^{-1}(r)\frac{d\rho(r)}{dr}$$

However, as pointed out by Crossley and Rochester (1980), one cannot retain it any more as defined by the original density gradient $d\rho/dr$ because this would make N^2 very strongly positive while the size of N^2 in the Earth's core is known to be very small (figure 1.25). A density gradient compatible with this fact almost vanishes in the case of incompressibility.

The change of density defined by eqn 3.12 vanishes ($\partial\rho/\partial t = 0$ when $\operatorname{div}\bar{q} = 0$ and $\operatorname{grad}\rho = 0$) and the Poisson equation reduces to $\nabla^2 V_1 = 0$, which makes V_1 harmonic throughout the core. Then, with a core comprised between rigid boundaries, V_1 will be of the form r^{-l-1} at the inner boundary and r^l at the outer boundary (Kelvin reciprocity theorem) and therefore vanishes inside the core. It results that the Poisson equation no longer plays a role (Crossley and Rochester, 1980).

Newtonian viscosity for an incompressible fluid

The Newton body is a liquid defined by the rheological law:

$$\begin{cases} p_m = ke_v + \eta_v \dot{e}_v \\ p_0 = 2\eta\dot{e}_0 \end{cases} \tag{3.20}$$

e is the deformation tensor (e_v cubic dilatation, e_0 deviator), p is the stress tensor with p_m the mean tension and p_0 the deviator. The rheological constants are:

k which takes care of the fact that under a purely hydrostatic pressure this liquid has the same behaviour as a Hooke solid;

η is the coefficient of shear dynamical viscosity;

η_v is the coefficient of bulk dynamic viscosity which role is to damp volumetric vibrations in case of a sudden variation of hydrostatic pressure. *This coefficient is not to be taken into account for an incompressible fluid.*

Stream function for an incompressible bidimensional fluid

For a great number of experimental situations the flow is bidimensional; for example a laminar flow in the plane (x, z) with $v = 0$, the flow in a boundary layer, or, in the case of a rigid rotation around the axis Oz, a flow independent of the longitude λ.* Let us suppose here that the velocity is independent of z. The continuity equation reduces to:

$$\frac{\partial u}{\partial x} + \frac{\partial v}{\partial y} = 0 \tag{3.21}$$

One immediately sees the advantage of introducing a *stream function* $\psi(x, y, t)$ such that:

$$u = \frac{\partial \psi}{\partial y} \qquad v = -\frac{\partial \psi}{\partial x} \tag{3.22}$$

A consequence is that

$$\text{curl}\,\bar{q} = \hat{k}\left(\frac{\partial v}{\partial x} - \frac{\partial u}{\partial y}\right) = \hat{k}\,\nabla_s^2 \psi = 2\bar{\omega}\,\dagger \tag{3.23}$$

The stream function has for dimensions $L^2 T^{-1}$: it is the surface crossed within 1 second. One could have, for example

$$\psi(x, y, t) = \phi(y)\,e^{i(\alpha x - \beta t)} \tag{3.24}$$

which represents an oscillation where $\lambda = 2\pi/\alpha$ is the wave length and β the frequency.

Spherical coordinates

It will be convenient to us to use such coordinates and write, according to eqn (3.18):

$$\text{div}\,\bar{q} = \frac{\partial u}{\partial r} + \frac{\partial v}{r\partial \theta} + \frac{\partial w}{r \sin\theta\, \partial \lambda} + 2\frac{u}{r} + \cot\theta\,\frac{v}{r} = 0 \tag{3.25}$$

* In that specific case cylindrical coordinates are more appropriate than spherical ones.

† $\nabla_s^2 = \nabla_H^2 = \nabla^2 - (\hat{k}\cdot\nabla)^2$ is called 'horizontal Laplacian". In spherical coordinates it is often written

$$L^2 = -\left[\frac{1}{\sin\theta}\frac{\partial}{\partial\theta}\left(\sin\theta\frac{\partial}{\partial\theta}\right) + \frac{1}{\sin^2\theta}\frac{\partial^2}{\partial\lambda^2}\right]$$

which is the angular part of the Laplacian. $(\hat{\imath}, \hat{\jmath}, \hat{k})$ are the unit vectors along the axes.

u, v, w being the components of the velocity along increasing r, θ, λ.

In a body rotating around OZ, the speed is independent from λ:

$$\frac{\partial w}{r \sin \theta \partial \lambda} = 0 \tag{3.26}$$

Consequently, one can put

$$\boxed{u = \frac{1}{r^2 \sin \theta} \frac{\partial \psi}{\partial \theta}, \qquad v = -\frac{1}{r \sin \theta} \frac{\partial \psi}{\partial r}} \tag{3.27}$$

We can check that

$$\left.\begin{aligned} \frac{\partial u}{\partial r} &= -\frac{2}{r^3 \sin \theta} \frac{\partial \psi}{\partial \theta} + \frac{1}{r^2 \sin \theta} \frac{\partial^2 \psi}{\partial \theta \partial r} \\ \frac{\partial v}{r \partial \theta} &= \frac{\cos \theta}{r^2 \sin^2 \theta} \frac{\partial \psi}{\partial r} - \frac{1}{r^2 \sin \theta} \frac{\partial^2 \psi}{\partial r \partial \theta} \\ 2\frac{u}{r} &= \frac{2}{r^3 \sin \theta} \frac{\partial \psi}{\partial \theta} \\ \cot \theta \frac{v}{r} &= -\frac{\cos \theta}{r^2 \sin^2 \theta} \frac{\partial \psi}{\partial r} \end{aligned}\right\} \tag{3.28}$$

Equation (3.25) is satisfied.

Cylindric coordinates (ρ, z, λ)

$$\rho = r \sin \theta \qquad z = r \cos \theta \ (\theta\text{: colatitude}) \tag{3.29}$$

One has

$$\operatorname{div} \bar{q} = \frac{1}{\rho} \frac{\partial}{\partial \rho} (\rho u) + \frac{\partial w}{\rho \partial \lambda} + \frac{\partial v}{\partial z} = 0 \tag{3.30}$$

or

$$\frac{\partial u}{\partial \rho} + \frac{u}{\rho} + \frac{\partial v}{\partial z} + \frac{\partial w}{\rho \partial \lambda} = 0 \tag{3.31}$$

If $\dfrac{\partial w}{\rho \partial \lambda} = 0$ one puts $\quad u = \dfrac{1}{\rho} \dfrac{\partial \psi}{\partial z} \quad v = -\dfrac{1}{\rho} \dfrac{\partial \psi}{\partial \rho}$ (3.32)

One has to give a convenient form to the component w which is independent of λ.

For a rigid solid rotation

$$w = \omega r \sin \theta = \omega \rho \tag{3.33}$$

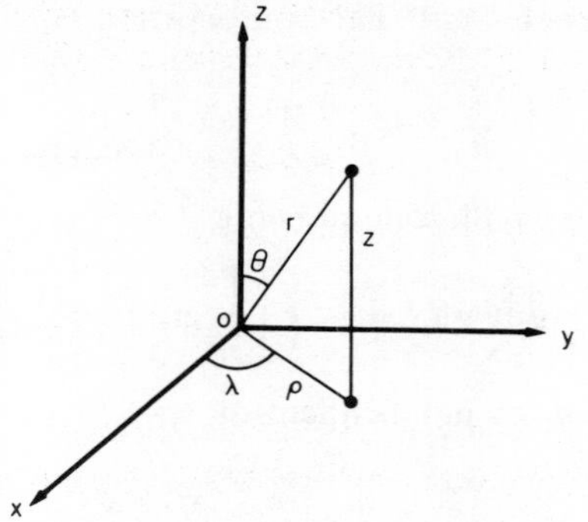

FIGURE 3.1.

so that several authors write:

$$\left.\begin{aligned} w &= \frac{\chi(r)}{\rho} = \frac{\chi(r)}{r\sin\theta} \\ \chi(r) &= \omega r^2 \sin^2\theta \end{aligned}\right\} \tag{3.34}$$

which fits eqn (3.33). It will then be easy to introduce small perturbations of speed by writing:

$$\chi(r) = \omega r^2 \sin^2\theta + \varepsilon a^2 \omega \chi(r) \tag{3.35}$$

Conservation of momentum: the Navier–Stokes equation

The process of internal friction takes place if the different regions of the fluid have different velocities, the result of friction being to make the speed uniform. The coefficient of kinematic viscosity is the resistance matter opposes to differences of velocities. Therefore the viscous stress tensor depends upon the derivatives of the velocity with respect to the coordinates and, *if the gradients are not too strong*, one can restrict the analysis to the first derivatives with a linear dependency:

$$\tau_{ik} = a\left(\frac{\partial q_i}{\partial x_k} + \frac{\partial q_k}{\partial x_i}\right) + b\frac{\partial q_l}{\partial x_l}\delta_{ik} \tag{3.36}$$

or, in a more convenient form (Landau and Lifschitz, Fluid Mechanics 1971).

$$\tau_{ik} = \eta\left(\frac{\partial q_i}{\partial x_k} + \frac{\partial q_k}{\partial x_i} - \frac{2}{3}\delta_{ik}\frac{\partial q_l}{\partial x_l}\right) + \eta_v \delta_{ik}\frac{\partial q_l}{\partial x_l} \tag{3.37}$$

the first parenthesis is zero when summing components such that $i = k$; η_v is the second viscosity coefficient. The conservation equations are

$$\rho\left(\frac{\partial q_i}{\partial t}+q_k\frac{\partial q_i}{\partial x_k}\right)=-\frac{\partial p}{\partial x_i}+\frac{\partial \tau_{ik}}{\partial x_k} \tag{3.38}$$

the force exerted on a small volume being

$$-\oiint_S p\,dS=-\iiint_V \operatorname{grad} p\,dV.$$

If we consider η and η_v as independent of x_i:

$$\rho\left(\frac{\partial \bar{q}}{\partial t}+(\bar{q}\cdot\operatorname{grad})\bar{q}\right)=-\operatorname{grad}P+\eta\nabla^2\bar{q}+\left(\eta_v+\frac{\eta}{3}\right)\operatorname{grad}\operatorname{div}\bar{q} \tag{3.39}$$

If the fluid is incompressible

$$\operatorname{div}\bar{q}=\nabla\cdot\bar{q}=0$$

we obtain the well-known *Navier–Stokes equation*

$$\boxed{\frac{\partial \bar{q}}{\partial t}+(\bar{q}\cdot\nabla)\bar{q}=-\nabla p+\nu\nabla^2\bar{q}} \tag{3.40}$$

with

$$\nu=\eta/\rho \tag{3.41}$$

The non-linear term $(\bar{q}\cdot\nabla)\bar{q}$ is called the *advection.* Equation (3.40) expresses the fact that a variation of the velocity $\bar{q}$ is to be ascribed to an unbalance between pressure forces and viscous forces.

Other forces will have to be considered in the following chapters, for example the Coriolis force in a rotating frame of reference, the Archimedes force for a stratified fluid, the Lorentz force in the presence of the magnetic field. They have to be added to achieve the balance of forces in eqn (3.40) but when the amplitudes of the oscillations can be assumed small enough, the equation can be linearized by dropping the advection term.

With linear equations, small oscillations can be resolved into a set of normal modes where the motions are harmonic and independent of other modes.

3.3 Dimensionless characteristic numbers in hydrodynamics

As the velocity field is solenoidal ($\nabla\cdot\bar{q}=0$), the Navier equation for a rotating fluid body can be written:

$$\frac{\partial \bar{q}}{\partial t}+(\bar{q}\cdot\nabla)\bar{q}+2\bar{\Omega}\wedge\bar{q}=-\nabla p-\nu\nabla\wedge(\nabla\wedge\bar{q}) \tag{3.42}$$

(p is now the reduced pressure).

The heat equation has a similar form:

$$\frac{\partial T}{\partial t}+(\bar{q}\cdot\nabla)T=\chi\nabla^2 T\left(+\frac{1}{\rho C_p}H\right) \tag{3.43}$$

where H is the heat generated by internal radioactive sources.

Let us introduce several dimensionless numbers:

$$E=\frac{\nu}{\Omega L^2}\therefore\frac{\text{viscosity effect}}{\text{Coriolis force}}\therefore\text{ Ekman number (three parameters)} \tag{3.44}$$

$$Ta=\frac{\Omega^2 L^4}{\nu^2}\therefore\left(\frac{\text{Coriolis force}}{\text{viscous force}}\right)^2\therefore\text{ Taylor number (three parameters)} \tag{3.45}$$

$$\varepsilon=\frac{UL}{\Omega L^2}=\frac{U}{\Omega L}\therefore\frac{\text{advection}}{\text{Coriolis force}}\therefore\text{ Rossby number (three parameters)} \tag{3.46}$$

$$P=\frac{\nu}{\chi}\therefore\frac{\text{viscous diffusion}}{\text{thermal diffusion}}\therefore\text{ Prandtl number (two parameters)} \tag{3.47}$$

and change the variables in (3.42) and (3.43) as follows:

$$\begin{aligned}
&\bar{r}\to L\bar{r} && \text{(in } \nabla \text{ operators)}\\
&t\to\Omega^{-1}t\\
&\bar{q}\to U\bar{q} && (U\text{: linear speed})\\
&\bar{\Omega}\to\Omega\hat{k}\\
&p\to\rho\Omega U L p
\end{aligned}$$

With such a scaling—which is equivalent to a change of the system of units—the system of fundamental equations becomes:

$$\boxed{\begin{aligned}
&\nabla\cdot\bar{q}=0\\
&\left(\frac{\partial}{\partial t}+\varepsilon\bar{q}\cdot\nabla\right)\bar{q}+2\hat{k}\Lambda\bar{q}=E\nabla^2\bar{q}-\nabla p\\
&\left(\frac{\partial}{\partial t}+\varepsilon\bar{q}\cdot\nabla\right)T \qquad =P^{-1}E\nabla^2 T
\end{aligned}} \tag{3.48}$$

It is also usual in hydrodynamics to introduce:

$$R=\frac{\varepsilon}{E}=\frac{\rho UL}{\eta}=\frac{UL}{\nu}\therefore\text{ Reynolds number (three parameters)} \tag{3.49}$$

which does not imply rotation and controls the establishment of turbulence, and

$$St = \frac{nL}{U} \therefore \text{ Strouhal number (three parameters)} \tag{3.50}$$

where n is the frequency of oscillation of the boundary.

Scaling procedure is essential to construct laboratory reduced models representing correctly—if possible—the phenomena occurring at large planetary scale.

As we will see later on, this is indeed possible for pure hydrodynamics but not for magnetohydrodynamics. A characteristic number with three parameters means that we can play with two parameters, the third one being then constrained by the numerical values given to the other two.

The dimensionless Rayleigh number will be introduced in §3.18.

Linearization of the Navier equations

As long as one considers *small* perturbations of an initial steady state which is usually the hydrostatic equilibrium, non-linear terms can be neglected. The linearization consists, of course, in dropping the quadratic term $\bar{q} \cdot \nabla \bar{q}$. The condition is that $\varepsilon < E$.*

There is no alternative, because we are unable to integrate the complete equations. One can justify this approximation when the Rossby number ε is small with respect to the other characteristic numbers. The linearization has some important consequences:

1. The reversibility of the secondary flows generated by the boundary layers, because quadratic terms have disappeared. As an example, if the mantle rotates slower than the inner core, the secondary flow goes from the pole towards the equator, and if the mantle rotates faster, the flow goes from the equator towards the pole. (see §3.19)
2. Advection produces a differential rotation; thus when we drop the quadratic terms we lose some possible characteristics of the flow, possibly resonances.
3. A dependence into $e^{im\lambda} \cdot e^{i\omega t}$ can be proposed, and an orthogonal set of solutions can be obtained in the form of spherical harmonics

$$v = V(r) r^{-1} P_l^m (\cos\theta) \exp(im\lambda + i\omega_j t) \tag{3.51}$$

* The Rossby number ε is very small in the Earth's core ($\sim 4 \times 10^{-7}$), which allows us to neglect the advection terms. However the Ekman number E is even much smaller (10^{-10} to 10^{-15}, probably).

3.4 Helmholtz equation

Let us consider first the Navier equation with its advection term but without rotation:

$$\frac{\partial \bar{q}}{\partial t} + (\bar{q}\cdot\nabla)\bar{q} = -\nabla p + \nu\nabla^2\bar{q}$$

the relation

$$\tfrac{1}{2}\nabla q^2 = \bar{q}\Lambda(\nabla\Lambda\bar{q}) + (\bar{q}\cdot\nabla)\bar{q}$$

allows us to write

$$\frac{\partial \bar{q}}{\partial t} - \bar{q}\Lambda(\nabla\Lambda\bar{q}) = -\nabla p - \frac{1}{2}\nabla q^2 + \nu\nabla^2\bar{q} \tag{3.52}$$

we put

$$\bar{\omega} = \nabla\Lambda\bar{q} \tag{3.53}$$

and we take the curl of (3.52) which makes the gradients disappear. We obtain the *Helmholtz equation*:

$$\boxed{\frac{\partial \bar{\omega}}{\partial t} = \nabla\Lambda(\bar{q}\Lambda\bar{\omega}) + \nu\nabla^2\bar{\omega}} \tag{3.54}$$

But

$$\nabla\Lambda(\bar{q}\Lambda\bar{\omega}) = \bar{\omega}\cdot\nabla\bar{q} - \bar{q}\cdot\nabla\bar{\omega} + \bar{q}\nabla\cdot\bar{\omega} - \bar{\omega}\nabla\cdot\bar{q}$$

the last two terms are zero so that

$$\frac{\partial \bar{\omega}}{\partial t} + (\bar{q}\cdot\nabla)\bar{\omega} = \bar{\omega}\cdot\nabla\bar{q} + \nu\nabla^2\bar{\omega} \tag{3.55}$$

or, finally

$$\frac{d\bar{\omega}}{dt} = \text{curl}\,\bar{q}\cdot\text{grad}\,\bar{q} + \nu\nabla^2\bar{\omega} \tag{3.56}$$

where the last term represents a creation of vorticity by a diffusive transport.

In rotating axes

$$\boxed{\frac{d\bar{\omega}}{dt} + 2\bar{\Omega}\Lambda\bar{\omega} = \text{curl}\,\bar{q}\cdot\text{grad}\,\bar{q} + \nu\nabla^2\bar{\omega}} \tag{3.57}$$

As $\bar{q}$ is a transport quantity its derivatives and its curl, are also transport quantities. The Helmholtz equation, which expresses the torque balance, is the transport equation of vorticity.

3.5 Geostrophic flow, Taylor columns

The consideration of geostrophic flow was introduced by Napier Shaw in 1916. It is an approximation which is valid when the Rossby number and the Ekman number are extremely small.

In that case the Coriolis force dominates and is balanced by pressure gradient only.

In the Earth's core the size of the different terms is:

for the advection: $\bar{q}\cdot\nabla\bar{q} \sim 10^{-10}$
for the viscosity: $\nu\nabla^2\bar{q} \sim 10^{-20}$
for the Coriolis force: $2\bar{\Omega}\Lambda\bar{q} \sim 10^{-5}$

Thus the conditions for geostrophic flow should apply so that eqn. (3.48) becomes

$$2\hat{k}\Lambda\bar{q} = -\nabla p \tag{3.58}$$

(the axis Oz being of course taken along $\bar{\Omega}$), while

$$\nabla\cdot\bar{q} = 0$$

and

$$\bar{q}\cdot\hat{n} = 0 \text{ on the boundary.}$$

Let us take the curl of (3.58):

$$\hat{k}\cdot\text{grad}\,\bar{q} = 0 \tag{3.59}$$

this means that $\bar{q}$ is independent of z so that any fluid column will move as a whole, preserving its height (the container should be of more or less total constant height).

We can see that important property by developing eqn (3.58) as follows:

$$\left\{\begin{aligned} -2v + \frac{\partial p}{\partial x} &= 0 && (3.60)\\ 2u + \frac{\partial p}{\partial y} &= 0 && (3.61)\\ \frac{\partial p}{\partial z} &= 0 && (3.62)\end{aligned}\right.$$

Deriving the first two relations with respect to z gives:

$$-2\frac{\partial v}{\partial z} + \frac{\partial}{\partial z}\left(\frac{\partial p}{\partial x}\right) = 0, \qquad 2\frac{\partial u}{\partial z} + \frac{\partial}{\partial y}\left(\frac{\partial p}{\partial z}\right) = 0$$

and by considering (3.62) it results that

$$\frac{\partial u}{\partial z} = 0 \qquad \frac{\partial v}{\partial z} = 0 \tag{3.63}$$

A cross-differentiation of (3.60) and (3.61) also gives

$$\frac{\partial v}{\partial y} + \frac{\partial u}{\partial x} = 0$$

and, because of (3.16)

$$\frac{\partial w}{\partial z} = 0 \tag{3.64}$$

thus the flow (u, v, w) does not depend on z; *it is bidimensional.*

Very interesting experiences in containers are shown in Busse papers. Such experiences must evidently be scaled according to the dimensionless characteristic numbers. This is indeed possible in the case of purely hydrodynamic motions because the kinematic viscosity of water is practically the same as the kinematic viscosity of the liquid core ($E = 10^{-15}$). Thus, to correctly scale the phenomena with containers which are obviously too small, it is necessary to increase their speed of rotation (about 8 turns per second). A schematic diagram, reproduced on figure 3.3, shows how the photographs of laboratory experiments of Busse and Carrigan (1976) look. It is important to point out that to simulate a gravity gradient towards the centre of the sphere, a temperature gradient has been introduced between the centre and the surface, which produces an artificial buoyancy. Moreover the curvature of the container pushes the fluid near the boundary towards the equator, as shown on figure 3.3 taken from Carrigan and Gubbins (1979).

It is not clear what the influence of a magnetic field will be on such Taylor columns. Laboratory models are not practicable because the decay of the magnetic field is extremely fast for small bodies and a "magnetic scaling" does not appear feasible (see chapter 4).

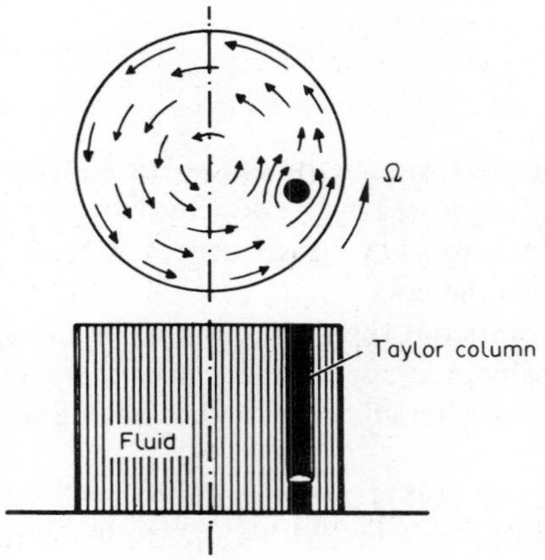

FIGURE 3.2 Generation of a Taylor column in a rotating tank (A. R. Robinson, 1978).

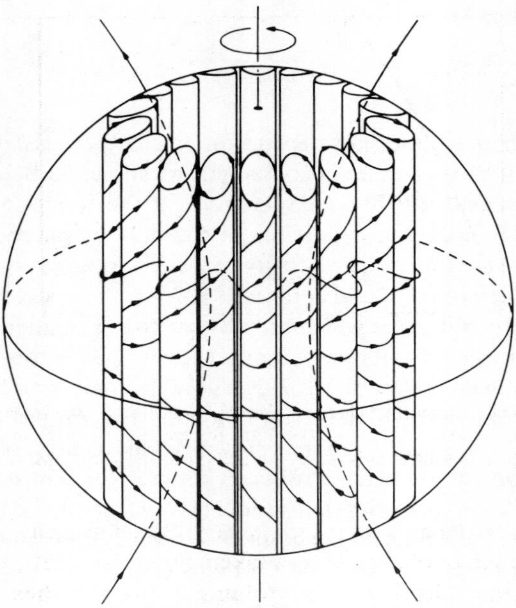

FIGURE 3.3 Convection circulations in the rotating spherical model have the appearance of slowly spinning rollers. Because the rollers end on the surface of a sphere, the fluid motions there are not parallel to the equatorial plane of the model. The boundary at each end of a roller is oppositely sloped, and so, to an observer of the model, outward circulation in a roller causes the fluid near the boundary to be pushed upward in the lower hemisphere and downward in the upper hemisphere. These motions are capable of generating the earth's dipolar magnetic field (according to Carrigan and Gubbins, 1979).

However, owing to the fast rotation of the Earth, an unsmooth core mantle interface like this one suggested by the downwards continuation of the gravity field should have quite important consequences on the distribution and size of Taylor columns inside the core.

Lighthill (1966) points out the distinction to be made between *thin sheets* and *fat bodies*. Geostrophic approximation is characteristic of thin sheets. Taylor columns are characteristic of fat bodies having similar upper and lower boundaries.

Smylie and Rochester (1981) question the reliability of extrapolating such laboratory demonstrations to the Earth's liquid core on the basis of the need to accommodate the flow in the boundary layers to a spherical container and the need to include compressibility and stratification.

3.6 Classical approximations

The Boussinesq approximation

Let us write eqn. (3.7) in the form

$$\rho \frac{D\bar{q}}{Dt} = -\nabla P + \rho \bar{g} \tag{3.65}$$

($\bar{g} = \nabla W$) and consider the hydrostatic equilibrium as the initial undisturbed state:

$$\nabla P_0 = \rho_0 \bar{g}, \quad \bar{g} = -g\hat{z}$$

One can expand P and ρ about the hydrostatic equilibrium values as:

$$\begin{aligned} P &= P_0 + P' \\ \rho &= \rho_0 + \rho' \end{aligned} \tag{3.66}$$

so that eqns (3.65) and (3.66) give:

$$\left(1 + \frac{\rho'}{\rho_0}\right)\frac{D\bar{q}}{Dt} = -\frac{1}{\rho_0}\nabla P' + \frac{\rho'}{\rho_0}\bar{g} \tag{3.67}$$

The *Boussinesq approximation* is a classical approximation in hydrodynamics which consists in neglecting ρ'/ρ_0 in the left-hand side where it produces a negligible correction to inertia but to keep it in the buoyancy term $(\rho'/\rho_0)\bar{g}$ where it is important.

The vertical equation of motion of a stratified Boussinesq fluid in a uniform gravitational field is thus

$$\frac{\partial w}{\partial t} = -\frac{\rho'}{\rho_0} g \tag{3.68}$$

while the incompressibility condition gives for the second equation (3.12):

$$\frac{\partial \rho'}{\partial t} = -w \frac{\partial \rho}{\partial z} \tag{3.69}$$

it results

$$\frac{\partial^2 \rho'}{\partial t^2} - \left(\frac{g}{\rho}\frac{\partial \rho}{\partial z}\right)\rho' = 0 \tag{3.70}$$

an equation governing oscillations with the Brunt–Väisälä frequency

$$N^2 = -\frac{g}{\rho}\frac{\partial \rho}{\partial z}$$

(here $k = \infty$).

The "traditional approximation" in meteorology and oceanography

The linearized Navier Stokes equations (3.8) can be written, in spherical coordinates, as follows (see Appendix, page 238):

$$\left.\begin{aligned} i\omega_j u - 2\Omega w \sin\theta &= -\frac{\partial p}{\partial r} \\ i\omega_j v - 2\Omega w \cos\theta &= -\frac{1}{r}\frac{\partial p}{\partial \theta} \\ i\omega_j w + 2\Omega v \cos\theta + 2\Omega u \sin\theta &= -\frac{\partial p}{r \sin\theta \partial\lambda} \end{aligned}\right\} \quad (3.71)$$

The variables cannot be separated unless an approximation called the "traditional approximation" is made, which consists in considering that only the vertical component of the Earth's rotation $\Omega \cos\theta$ is important, and that the horizontal component $\Omega \sin\theta$ can be neglected. This approximation was used by Laplace in 1832. It is valid for sheets which are thin with respect to the Earth's radius, such as the oceans and the atmosphere, but also for strongly stratified liquids which can be considered as made of the superposition of several thin shells of homogeneous hydrostatic fluids.

However, the horizontal component cannot be neglected for internal (buoyancy) waves in weakly stratified deep layers (Needler and LeBlond, 1973).

The β plane approximation (*Rossby*, 1939)

Rossby has overcome the mathematical difficulties related with the need to work with spherical coordinates in meteorology and oceanography by using a set of plane coordinates which retains only the major effect of the Coriolis force: the spherical surface is approximated locally by a plane surface.

The basic idea is that for a displacement towards north there will be a force acting southwards, due to an increase of the Coriolis parameter

$$f = 2\Omega \sin\phi \quad (3.72)$$

However, this approximation should be valid only for *thin rotating spherical shells* and for regions remote from the equator: at mean latitudes one can approximate the spherical surface with a plane where one admits that the northward derivative

$$\beta = \frac{\partial f}{\partial \phi} = 2\Omega \cos\phi = \mathit{constant} \quad (3.73)$$

In that plane one chooses as a local Cartesian frame an axis Oy towards the north ($y \equiv \phi$) and an axis Ox towards the east while $\Omega \sin\phi$, the vertical component of Ω, is taken as the local angular velocity of rotation.

This eliminates most of the mathematical difficulties and allowed Rossby to demonstrate the existence of a class of large-scale *inertial toroidal waves* (no radial component) called planetary waves or Rossby waves.

As div $\bar{q} = 0$, we can introduce in the β plane ($\bar{q} = \bar{q}(u, v)$) a stream function $\psi(x, y, t)$ such that

$$u = \frac{\partial \psi}{\partial y}, \quad v = -\frac{\partial \psi}{\partial x} \tag{3.74}$$

then the Navier equations become

$$\left.\begin{aligned} \frac{d}{dt}\frac{\partial \psi}{\partial y} + f\frac{\partial \psi}{\partial x} &= -\frac{1}{\rho}\frac{\partial p}{\partial x} \\ -\frac{d}{dt}\frac{\partial \psi}{\partial x} + f\frac{\partial \psi}{\partial y} &= -\frac{1}{\rho}\frac{\partial p}{\partial y} \end{aligned}\right\} \tag{3.75}$$

and taking $\frac{\partial}{\partial y}$ (of the first) $-\frac{\partial}{\partial x}$ (of the second) gives

$$\frac{d}{dt}(\nabla_H^2 \psi) + \beta \frac{\partial \psi}{\partial x} = 0 \tag{3.76}$$

$$\left(\nabla_H^2 = \frac{\partial^2}{\partial x^2} + \frac{\partial^2}{\partial y^2}\right) \tag{3.77}$$

For *small* perturbations of the permanent state we can linearize these equations and seek solutions in the form of *plane waves:*

$$\psi = e^{i(ax + by - \sigma t)} \tag{3.78}$$

of wave lengths $2\pi/a$ and $2\pi/b$ in the x and y directions respectively and with angular frequency σ in time.

Moreover, introducing the solution (3.78) into (3.76) gives

$$\sigma(a^2 + b^2) + \beta a = 0 \tag{3.79}$$

A similar equation is easily obtained for the vorticity

$$\xi = \frac{\partial v}{\partial x} - \frac{\partial u}{\partial y}$$

These oscillations are called *inertial* because they are dominated by the Coriolis force. These waves are dispersive, their phase velocity is:

$$\frac{\sigma}{a} = -\frac{\beta}{a^2 + b^2} = -\frac{\beta}{k^2} \tag{3.80}$$

while their group velocities are:

$$\frac{\partial\sigma}{\partial a}=\beta\frac{a^2-b^2}{(a^2+b^2)^2}=\frac{\beta}{k^2}\cos 2\alpha$$

$$\frac{\partial\sigma}{\partial b}=\beta\frac{2ab}{(a^2+b^2)^2}=\frac{\beta}{k^2}\sin 2\alpha \tag{3.81}$$

this shows that the group velocity vector makes an angle 2α with the Ox axis. The same equation has been obtained by Longuet–Higgins (1964) for a *thin* shell in between two synchronous rotating spheres where ψ obeys the equation:

$$\frac{d}{dt}(\nabla_S^2\psi)+2\Omega\frac{\partial\psi}{\partial\lambda}=0 \tag{3.82}$$

The β plane approximation—which is not applicable at the equator—is used to investigate plane wave oscillations in limited water basins on the surface of the Earth (like inland seas or lakes) and may be extended to *thin* shells, giving inertial modes of long period, greater than $2\pi/\Omega$ (Rossby or planetary waves). It cannot be applied to fat bodies like the Earth's core without much care. Hide (1966) has developed such investigations.

3.7 Differential equations of the perturbed pressure in a rotating spheroid

A pressure gradient is necessary to sustain an oscillatory mode. This justifies the formulation of the problem in terms of the pressure alone. (Friedlander and Siegmann, 1982).

The hydrostatic equilibrium is usually taken as the initial undisturbed state and one considers that small deviations from this initial state are produced by some event. The assumption of small disturbances permits the linearization of the Navier Stokes equation.

Viscous incompressible spheroid

Let us operate the following transformations upon the two components of the Navier equations:

$$\left.\begin{aligned}\frac{\partial\dot{u}}{\partial x}-2\Omega\frac{\partial v}{\partial x}&=-\frac{\partial^2 p}{\partial x^2}+\nu\nabla^2\frac{\partial u}{\partial x}\\ \frac{\partial\dot{v}}{\partial y}+2\Omega\frac{\partial u}{\partial y}&=-\frac{\partial^2 p}{\partial y^2}+\nu\nabla^2\frac{\partial v}{\partial y}\end{aligned}\right\} \tag{3.83}$$

or

$$\left. \begin{aligned} \left(\nu\nabla^2 - \frac{\partial}{\partial t}\right)\frac{\partial u}{\partial x} &= \frac{\partial^2 p}{\partial x^2} - 2\Omega\frac{\partial v}{\partial x} \\ \left(\nu\nabla^2 - \frac{\partial}{\partial t}\right)\frac{\partial v}{\partial y} &= \frac{\partial^2 p}{\partial y^2} + 2\Omega\frac{\partial u}{\partial y} \end{aligned} \right\} \tag{3.84}$$

with a solenoidal field of velocity

$$\frac{\partial u}{\partial x} + \frac{\partial v}{\partial y} + \frac{\partial w}{\partial z} = 0$$

this becomes

$$\left(\nu\nabla^2 - \frac{\partial}{\partial t}\right)\frac{\partial w}{\partial z} = -\left(\frac{\partial^2 p}{\partial x^2} + \frac{\partial^2 p}{\partial y^2}\right) + 2\Omega\left(\frac{\partial v}{\partial x} - \frac{\partial u}{\partial y}\right) \tag{3.85}$$

to which we apply once more the operator $\left(\nu\nabla^2 - \dfrac{\partial}{\partial t}\right)$:

$$\left(\nu\nabla^2 - \frac{\partial}{\partial t}\right)^2 \frac{\partial w}{\partial z} = -\left(\nu\nabla^2 - \frac{\partial}{\partial t}\right)\left(\frac{\partial^2 p}{\partial x^2} + \frac{\partial^2 p}{\partial y^2}\right)$$

$$+ 2\Omega\left\{\underbrace{\frac{\partial}{\partial x}\left(\nu\nabla^2 - \frac{\partial}{\partial t}\right)v}_{2\Omega u + \frac{\partial p}{\partial y}} - \underbrace{\frac{\partial}{\partial y}\left(\nu\nabla^2 - \frac{\partial}{\partial t}\right)u}_{-2\Omega v + \frac{\partial p}{\partial x}}\right\}$$

$$4\Omega^2\left(\frac{\partial u}{\partial x} + \frac{\partial v}{\partial y}\right) = -4\Omega^2\frac{\partial w}{\partial z}$$

or

$$\left\{\left(\nu\nabla^2 - \frac{\partial}{\partial t}\right)^2 + 4\Omega^2\right\}\frac{\partial w}{\partial z} = -\left(\nu\nabla^2 - \frac{\partial}{\partial t}\right)\left(\frac{\partial^2 p}{\partial x^2} + \frac{\partial^2 p}{\partial y^2}\right) \tag{3.86}$$

Let us introduce now the z derivative of the third Navier equation:

$$\left(\nu\nabla^2 - \frac{\partial}{\partial t}\right)\frac{\partial w}{\partial z} = \frac{\partial^2 p}{\partial z^2} \tag{3.87}$$

eqn (3.86) gives

$$\left(\nu\nabla^2 - \frac{\partial}{\partial t}\right)\nabla^2 p + 4\Omega^2\frac{\partial w}{\partial z} = 0 \tag{3.88}$$

and we apply once more the operator $\left(\nu\nabla^2 - \frac{\partial}{\partial t}\right)$:

$$\left(\nu\nabla^2 - \frac{\partial}{\partial t}\right)^2 \nabla^2 p + 4\Omega^2\left(\nu\nabla^2 - \frac{\partial}{\partial t}\right)\frac{\partial w}{\partial z} = 0 \tag{3.89}$$

which, by using (3.87) gives the *Greenspan equation:*

$$\boxed{\left(\nu\nabla^2 - \frac{\partial}{\partial t}\right)^2 \nabla^2 p + 4\Omega^2 \frac{\partial^2 p}{\partial z^2} = 0} \tag{3.90}$$

If one multiplies both sides by $i^2 = -1$ one has

$$\left(i\nu\nabla^2 - i\frac{\partial}{\partial t}\right)^2 \nabla^2 p = 4\Omega^2 \frac{\partial^2 p}{\partial z^2} \tag{3.91}$$

and, seeking for a solution $p = p_0 \exp(i\omega_i t)$

$$(i\nu\nabla^2 + \omega_i)^2 \nabla^2 p = 4\Omega^2 \frac{\partial^2 p}{\partial z^2} \tag{3.92}$$

a form proposed by Jeffreys (1949). In the case of a steady motion the Greenspan equation becomes

$$\left(E^2\nabla^6 + 4\frac{\partial^2}{\partial z^2}\right)p = 0 \tag{3.93}$$

with $E = \nu/\Omega L^2$ the Ekman number.

Elastic incompressible spheroid

When treating the problem for a Hooke body we have simply to replace the operator $\nu\partial/\partial t$ by an operator μ, μ being the rigidity coefficient and we obtain the *Hough equation:*

$$\boxed{\left(\mu\nabla^2 - \frac{\partial^2}{\partial t^2}\right)^2 \nabla^2 p + 4\Omega^2 \frac{\partial^2}{\partial t^2}\frac{\partial^2 p}{\partial z^2} = 0} \tag{3.94}$$

Perfect incompressible fluid spheroid

When $\mu = 0$, $\nu = 0$ the equation becomes

$$4\Omega^2 \frac{\partial^2 p}{\partial z^2} + \frac{\partial^2}{\partial t^2}\nabla^2 p = 0 \tag{3.95}$$

or

$$\frac{\partial^2}{\partial t^2}\left(\frac{\partial^2 p}{\partial x^2}+\frac{\partial^2 p}{\partial y^2}\right) = -\left(\frac{\partial^2}{\partial t^2}+4\Omega^2\right)\frac{\partial^2 p}{\partial z^2}$$

and, with $p = p_0 \exp(i\omega_i t)$

$$\omega_i^2\left(\frac{\partial^2 p}{\partial x^2}+\frac{\partial^2 p}{\partial y^2}\right) = (-\omega_i^2+4\Omega^2)\frac{\partial^2 p}{\partial z^2} \tag{3.96}$$

this is the *Poincaré equation* that we write as follows:

$$\boxed{\nabla^2 p - \frac{4\Omega^2}{\omega_i^2}\frac{\partial^2 p}{\partial z^2} = 0} \tag{3.97}$$

This second-order partial differential equation is of mixed type. If

$$1-\frac{4\Omega^2}{\omega_i^2} < 0, \text{ that is } -2\Omega < \omega_i < +2\Omega \tag{3.98}$$

the equation is an *hyperbolic differential equation* which is the typical form of wave propagation equations: small perturbations of an equilibrium configuration can propagate in the fluid in the form of waves which are said to be "*inertial*" because they are controlled by the Coriolis force as a restoring force. These are low-frequency modes. If

$$\omega_i > 2\Omega \tag{3.99}$$

it is an *elliptical differential equation* and non-zero solutions are impossible. The system does not differ much from a non rotating system for which $\nabla^2 p = 0$ (as $\Omega = 0$). One usually puts

$$\lambda = \frac{\omega_i}{\Omega} \tag{3.100}$$

Thus, in a fluid in uniform rotation, inertial oscillations could exist with a frequency lower than twice the rotation frequency of the fluid (i.e. periods greater than half a day in the case of the Earth's liquid core).

The Poincaré equation is typical for the conditions inside the liquid core.

The difficulties come from the fact that, if the boundary conditions are prescribed values of $\bar{q}$ on one or two spherical boundaries ($\bar{q}\cdot\hat{n} = 0$), this is an ill-posed problem in the sense of Hadamard.

When the boundary condition is of the generalized Dirichlet type, i.e. a relation between the function p and its normal derivative, this is appropriate to an elliptic but not to an hyperbolic differential equation. There are no existence and uniqueness theorems for hyperbolic equations in that case (Stewartson and Roberts, 1963) so that the existence of continuous solutions appears very doubtful (Aldridge, 1972).

However, particular examples are evidence that solutions exist for a *discrete* set of values of λ but the solutions may not always be smooth.

Aldridge and Toomre (1969) and Aldridge (1975) have shown *experimentally* that solutions exist under certain conditions (discrete Eigen modes) (figure 3.4).

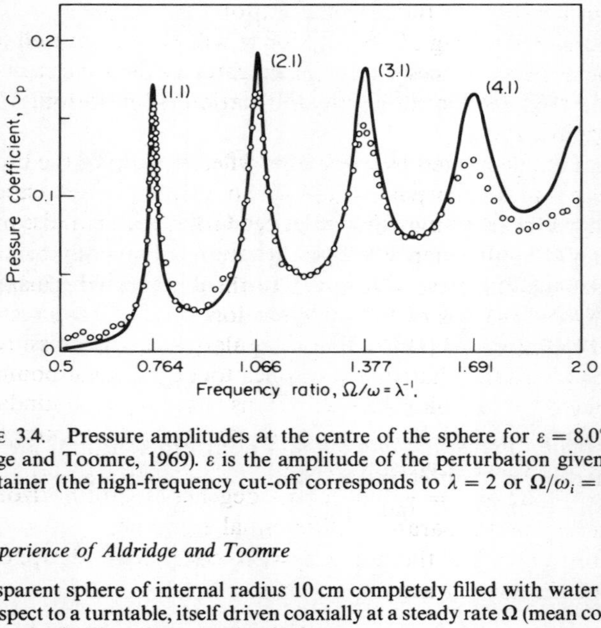

FIGURE 3.4. Pressure amplitudes at the centre of the sphere for $\varepsilon = 8.0°$ (from Aldridge and Toomre, 1969). ε is the amplitude of the perturbation given to the container (the high-frequency cut-off corresponds to $\lambda = 2$ or $\Omega/\omega_i = 0.5$).

The experience of Aldridge and Toomre

A transparent sphere of internal radius 10 cm completely filled with water rotates with respect to a turntable, itself driven coaxially at a steady rate Ω (mean container rotation speed). The sphere rotation is thus:

$$\Omega_s(t) = \Omega + \varepsilon\omega \cos \omega t$$

with $\omega \approx 6.2 \text{ rad s}^{-1} \approx 352° \text{ s}^{-1}$.

The ratio Ω/ω is changed by changing the speed of rotation Ω of the turntable. Resonant oscillations are measured through differences of pressure between the centre and the surface of the sphere. Below $\Omega/\omega_i = 0.5$ ($\lambda = 2$) practically no fluid response was detected, as expected from theory (ω_i high-frequency cut-off).

The solid curve is theoretical. All observed resonances involve successive members of the $(n, 1)$ series of modes: with the system of measurements no modes other than those of the $(n, 1)$ series were observed, which reflects the fact that measurements of the centre to pole pressure differences discriminates against modes having more than one layer of cells in the vertical direction and especially those for which that number is even. By changing the measurement point position several $(n,2)$ modes were observed. One should note also that the lines are broadened by the small viscosity of water (Ekman number $\sim 1.5 \times 10^{-5}$ in this experiment).

If we write the Poincaré equation under the form:

$$\frac{\partial^2 p}{\partial x^2}+\frac{\partial^2 p}{\partial y^2}-\frac{1}{\lambda^2}(4-\lambda^2)\frac{\partial^2 p}{\partial z^2}=0 \tag{3.101}$$

its characteristic surfaces are real and are dubble revolution cones

$$(x^2+y^2)^{1/2} \pm \lambda(4-\lambda^2)^{-1/2}\, z = c^{ste} \tag{3.102}$$

of semi-angle

$$\phi = \text{arc cot}\ \lambda^{-1}(4-\lambda^2)^{1/2} = \text{arc sin}\frac{\lambda}{2} = \text{arc sin}\frac{\omega_i}{2\Omega} \tag{3.103}$$

symmetric with respect to the axis of rotation.

Disturbances originating at the boundary will be transmitted along these conical characteristic surfaces: energy propagates along wave crests which are inclined with angle ϕ on the axis of rotation (Bretherton, Carrier and Longuet–Higgins, 1966).

Other cones are generated by successive reflections upon the boundaries—see photograph in Greenspan (1969)—who demonstrated experimentally these shear cones in a rotating cylinder (speed Ω) having a flat disc inside which is oscillating with ω_i frequency.

Such surfaces should have a thickness of the order of $E^{1/3}$ inside the fluid. They are singular surfaces of the wave motion.

Stewartson and Rickard (1969) found singularities in the pressure at the two critical circles where the characteristic cones touch the shell boundaries. The fluid could develop "pathological oscillations" near to the boundary at these critical latitudes where non linear terms are essential (weak spots: Greenspan, 1964, p. 687; Bondi and Lyttleton, 1953).

When $\omega_i = \pm 2\Omega$ or $\lambda = \pm 2$ the cones degenerate into horizontal planes: the equation becomes a parabolic differential equation.

Finally, for $|\omega_i| > 2\Omega$ the wave system completely disappears as the equation is of elliptic character.

Compressible perfect fluid spheroid

We consider in that case the displacements $(U\,V\,W)$ instead of their velocities so that the Navier equations are

$$\left.\begin{aligned} \ddot{U} - 2\Omega\dot{V} &= -\frac{\partial p}{\partial x} \\ \ddot{V} + 2\Omega\dot{U} &= -\frac{\partial p}{\partial y} \\ \ddot{W} &= -\frac{\partial p}{\partial z} \end{aligned}\right\} \tag{3.104}$$

and with an exp $(i\omega_i t)$ as solution

$$\left.\begin{aligned} -\omega_i^2 U - 2\Omega\dot{V} &= -\frac{\partial p}{\partial x} \\ -\omega_i^2 V + 2\Omega\dot{U} &= -\frac{\partial p}{\partial y} \\ -\omega_i^2 W \qquad &= -\frac{\partial p}{\partial z} \end{aligned}\right\} \quad \begin{aligned} -\omega_i^2 \dot{V} + 2\Omega\dot{U} &= -\frac{\partial \dot{p}}{\partial y} \\ \\ -\omega_i^2 U - 2\Omega\left(\frac{2\Omega}{\omega_i^2}\dot{U} + \frac{1}{\omega_i^2}\frac{\partial \dot{p}}{\partial y}\right) &= -\frac{\partial p}{\partial x} \end{aligned}$$

thus

$$\left.\begin{aligned} U(\omega_i^2 - 4\Omega^2) &= \frac{\partial p}{\partial x} - \frac{2\Omega}{\omega_i^2}\frac{\partial \dot{p}}{\partial y} \\ V(\omega_i^2 - 4\Omega^2) &= \frac{\partial p}{\partial y} + \frac{2\Omega}{\omega_i^2}\frac{\partial \dot{p}}{\partial x} \\ W\omega_i^2 &= \frac{\partial p}{\partial z} \end{aligned}\right\} \tag{3.105}$$

and, by derivation

$$\left.\begin{aligned} \frac{\partial U}{\partial x}(\omega_i^2 - 4\Omega^2) &= \frac{\partial^2 p}{\partial x^2} - \frac{2\Omega}{\omega_i^2}\frac{\partial^2 \dot{p}}{\partial y \partial x} \\ \frac{\partial V}{\partial y}(\omega_i^2 - 4\Omega^2) &= \frac{\partial^2 p}{\partial y^2} + \frac{2\Omega}{\omega_i^2}\frac{\partial^2 \dot{p}}{\partial x \partial y} \\ \frac{\partial W}{\partial z}(\omega_i^2 - 4\Omega^2) + 4\Omega^2\frac{\partial W}{\partial z} &= \frac{\partial^2 p}{\partial z^2} \end{aligned}\right\} \tag{3.106}$$

Summing:

$$(\omega_i^2 - 4\Omega^2)\Theta + 4\Omega^2\frac{\partial W}{\partial z} = \nabla^2 p \tag{3.107}$$

$$\Theta = \frac{\partial U}{\partial x} + \frac{\partial V}{\partial y} + \frac{\partial W}{\partial z} \quad \text{the cubical dilatation.}$$

This can be written:

$$\boxed{(\omega_i^2 - 4\Omega^2)\Theta = \nabla^2 p - \frac{4\Omega^2}{\omega_i^2}\frac{\partial^2 p}{\partial z^2}} \tag{3.108}$$

which gives the Poincaré equation when $\Theta = 0$.

Note again that in an incompressible ($\Theta = 0$) non-rotating ($\Omega = 0$) body the Poincaré equation becomes $\nabla^2 p = 0$, an elliptic differential equation which has no solution in the form of progressive waves. *It is the rotation which makes the equation hyperbolic.*

3.8 Boundary condition for the reduced pressure

In the linear theory for inertial waves in an inviscid fluid, we put as boundary condition:

$$\bar{q}\cdot\hat{n} = 0 \tag{3.109}$$

with $\lambda = \omega_i/\Omega$, the Navier equation is:

$$i\lambda\bar{q} + 2(\hat{k}\wedge\bar{q}) = -\nabla p \tag{3.110}$$

Let us take the scalar product with

$$(2\hat{k}\wedge\hat{n}):$$

$$i\lambda(2\hat{k}\wedge\hat{n})\cdot\bar{q} + (2\hat{k}\wedge\hat{n})\cdot(2\hat{k}\wedge\bar{q}) = -(2\hat{k}\wedge\hat{n})\cdot\nabla p$$

If $\bar{q}\cdot\hat{n} = 0$:

$$(2\hat{k}\wedge\bar{q})\cdot\hat{n} = -\hat{n}\cdot\nabla p$$

or

$$(2\hat{k}\wedge\hat{n})\cdot\bar{q} = +\hat{n}\cdot\nabla p$$

thus

$$i\lambda\hat{n}\cdot\nabla p + (2\hat{k}\wedge\hat{n})\cdot(2\hat{k}\wedge\bar{q}) + 2(\hat{k}\wedge\hat{n})\cdot\nabla p = 0$$

Let us multiply by $i\lambda$:

$$-\lambda^2\hat{n}\cdot\nabla p + i\lambda(2\hat{k}\wedge\hat{n})\cdot(2\hat{k}\wedge\bar{q}) + 2i\lambda(\hat{k}\wedge\hat{n})\cdot\nabla p = 0$$

We also have:

$$(\hat{k}\wedge\hat{n})\cdot(\hat{k}\wedge\hat{q}) = (\hat{k}\cdot\hat{k})\underbrace{(\hat{n}\cdot\bar{q})}_{0} - (\hat{k}\cdot\bar{q})(\hat{n}\cdot\hat{k})$$

$$= -(\hat{k}\cdot\bar{q})(\hat{n}\cdot\hat{k})$$

thus

$$-\lambda^2\hat{n}\cdot\nabla p - 4(\hat{k}\cdot i\lambda\bar{q})(\hat{n}\cdot\hat{k}) + 2i\lambda\,(\hat{k}\wedge\hat{n})\cdot\nabla p = 0$$

We take $i\lambda\bar{q}$ in the Navier equation and as $\hat{k}\wedge\bar{q}$ is orthogonal to $\hat{k}$, one has $\hat{k}\cdot(\bar{k}\wedge\bar{q}) = 0$, from which the condition:

$$-\lambda^2\hat{n}\cdot\nabla p + 4(\hat{n}\cdot\hat{k})(\hat{k}\cdot\nabla p) + 2i\lambda(\hat{k}\wedge\hat{n})\cdot\nabla p = 0 \tag{3.111}$$

as given by Greenspan (1965) (eqn 1.3.8, p. 10; eqn 2.7.4, p. 51).

3.9 Solution of the Poincaré equation by separation of variables

Let us use cylindrical coordinates (ρ, z, Λ) and as we consider small-amplitude movements, we linearize the equations and seek for a plane waves solution:

$$p = p(\rho, z)\,e^{im\Lambda}\,e^{i\omega_i t} \tag{3.112}$$

m is the *azimuthal wave number*. We have:

$$\hat{k}\Lambda\hat{n} = \hat{\Lambda} \qquad \hat{n}\cdot\hat{k} = \cos\theta = \frac{z}{\sqrt{\rho^2+z^2}} \quad \frac{\partial p}{\partial \Lambda} = imp$$

and, consequently:

$$-\lambda^2(\hat{n}\cdot\nabla p) = -\lambda^2\left(\rho\frac{\partial p}{\partial \rho}+z\frac{\partial p}{\partial z}\right)\cdot\frac{1}{\sqrt{\rho^2+z^2}}$$

$$4(\hat{n}\cdot\hat{k})(\hat{k}\cdot\nabla p) = 4z\frac{\partial p}{\partial z}\cdot\frac{1}{\sqrt{\rho^2+z^2}}$$

$$2i\lambda(\hat{k}\Lambda\hat{n})\cdot\nabla p = 2i\,\lambda\frac{\partial p}{\partial \Lambda} = -2\lambda mp$$

Then the Poincaré equation becomes:

$$\frac{1}{\rho}\frac{\partial}{\partial \rho}\left(\rho\frac{\partial p}{\partial \rho}\right)-\frac{m^2}{\rho^2}p+\left(1-\frac{4}{\lambda^2}\right)\frac{\partial^2 p}{\partial z^2} = 0 \tag{3.113}$$

with the boundary condition

$$\frac{1}{\rho}\frac{\partial p}{\partial \rho}+\frac{2m}{\lambda}p+\left(1-\frac{4}{\lambda^2}\right)z\frac{\partial p}{\partial z} = 0 \tag{3.114}$$

on the sphere

$$r^2 = \rho^2+z^2 = 1 \tag{3.115}$$

Solutions of these relations represent inviscid inertial oscillations in the sphere.

If $m = 0$ (zero azimuthal wave number):

$$p = p(\rho, z)e^{i\omega_i t}; \quad U(r, z) = \frac{\partial p}{\rho\partial z}e^{i\omega_i t}; \quad W(r, z) = \frac{\partial p}{\rho\partial \rho}e^{i\omega_i t} \tag{3.116}$$

the Poincaré equation becomes

$$\frac{1}{\rho}\frac{\partial}{\partial \rho}\left(\rho\frac{\partial p}{\partial \rho}\right)-\gamma^2\frac{\partial^2 p}{\partial z^2} = 0 \tag{3.117}$$

with

$$\gamma^2 = -\left(1-\frac{4}{\lambda^2}\right) = \frac{4\Omega^2}{\omega_i^2}-1 \tag{3.118}$$

and

$$p = 0 \tag{3.119}$$

at the boundary (Aldridge, 1975).

3.10 Greenspan equation in cylindrical coordinates

Considering that the velocity field and pressure are independent of the longitude, Greenspan proposes a solution

$$\bar{q} = -\nabla\Lambda(\rho\chi(z,t)\hat{\Lambda}) + \rho V(z,t)\hat{\Lambda} \tag{3.120}$$

which eliminates the radial coordinate. The first term gives:

$$\left.\begin{aligned} q_\rho^{(1)} &= -2\omega_\rho = -\left(\frac{\partial u_z}{\rho\partial\Lambda} - \frac{\partial u_\lambda}{\partial z}\right) = +\rho\frac{\partial\chi}{\partial z} \\ q_\lambda^{(1)} &= -2\omega_\lambda = -\left(\frac{\partial u_\rho}{\partial z} - \frac{\partial u_z}{\partial\rho}\right) = 0 \\ q_z^{(1)} &= -2\omega_z = -\frac{1}{\rho}\left(\frac{\partial}{\partial\rho}(\rho u_\lambda) - \frac{\partial u_\rho}{\partial\lambda}\right) = -2\chi \end{aligned}\right\} \tag{3.121}$$

while the second one gives:

$$q_\rho^{(2)} = 0 \qquad q_\lambda^{(2)} = \rho V(z,t) \qquad q_z^{(2)} = 0 \tag{3.122}$$

so that the Coriolis force may be written:

$$2\hat{k}\Lambda(q_\rho\hat{\rho} + q_\lambda\hat{\Lambda} + q_z\hat{k}) = -2\left(\rho\frac{\partial\chi}{\partial z}\right)\hat{\Lambda} + 2\left(\rho V\right)\hat{\rho} \tag{3.123}$$

and the Navier equation

$$\left(E\nabla^2 - \frac{\partial}{\partial t}\right)\bar{q} - 2\hat{k}\Lambda\bar{q} = -\nabla p \tag{3.124}$$

then gives

$$\left.\begin{aligned} &\text{along } \hat{\Lambda} \quad \left(E\nabla^2 - \frac{\partial}{\partial t}\right)V - 2\frac{\partial\chi}{\partial z} = 0 \\ &\text{along } \hat{\rho} \quad \left(E\nabla^2 - \frac{\partial}{\partial t}\right)\frac{\partial\chi}{\partial z} + 2V = -\nabla p \end{aligned}\right\} \tag{3.125}$$

$\rho V(z,t)$ is an azimuthal tangential velocity

$\rho\chi(z,t)$ is a stream function.

Inside the boundary layer $\nabla p = 0$ (see p. 131).

Note: Because of obvious mathematical difficulties many theoretical investigations are conducted for the simplest possible container that is an unbounded space between two infinite parallel plates. It is true that this simplest case exhibits the basic hydrodynamic phenomena.

3.11 A Poincare equation for the velocity field

The three Navier equations for a perfect fluid:

$$\left.\begin{aligned} \dot{u}-2\Omega v &= -\frac{\partial p}{\partial x} \\ \dot{v}+2\Omega u &= -\frac{\partial p}{\partial y} \\ \dot{w} &= -\frac{\partial p}{\partial z} \end{aligned}\right\}$$

also give

$$\left.\begin{aligned} \frac{\partial}{\partial t}\left(\frac{\partial w}{\partial y}-\frac{\partial v}{\partial z}\right) &= 2\Omega\frac{\partial u}{\partial z} \\ \frac{\partial}{\partial t}\left(\frac{\partial u}{\partial z}-\frac{\partial w}{\partial x}\right) &= 2\Omega\frac{\partial v}{\partial z} \\ \frac{\partial}{\partial t}\left(\frac{\partial v}{\partial x}-\frac{\partial u}{\partial y}\right) &= -2\Omega\left(\frac{\partial u}{\partial x}+\frac{\partial v}{\partial y}\right) = 2\Omega\frac{\partial w}{\partial z} \end{aligned}\right\} \tag{3.126}$$

that is

$$\frac{\partial}{\partial t}\operatorname{curl}\bar{q} = 2\Omega\frac{\partial \bar{q}}{\partial z} \tag{3.127}$$

an equation governing the vorticity variations. Let us apply the operator $\partial/\partial t$ curl:

$$\frac{\partial^2}{\partial t^2}\operatorname{curl}\operatorname{curl}\bar{q} = -\frac{\partial^2}{\partial t^2}\nabla^2\bar{q} \quad \text{as } \nabla\cdot\bar{q}=0 \tag{3.128}$$

and

$$\frac{\partial}{\partial t}\operatorname{curl}\left(2\Omega\frac{\partial\bar{q}}{\partial z}\right) = 2\Omega\frac{\partial}{\partial z}\left(\frac{\partial}{\partial t}\operatorname{curl}\bar{q}\right) = 4\Omega^2\frac{\partial^2\bar{q}}{\partial z^2} \tag{3.129}$$

and we find a Greenspan equation in $\bar{q}$:

$$-\frac{\partial^2}{\partial t^2}\nabla^2\bar{q} = 4\Omega^2\frac{\partial^2\bar{q}}{\partial z^2} \tag{3.130}$$

where $v = 0$. It accepts plane wave solutions in the form

$$\bar{q} = \bar{q}_0 \exp[i(-\omega_i t + lx + my + nz)] \tag{3.131}$$

if

$$\omega_i^2(l^2+m^2+n^2) = 4\Omega^2 n^2 \tag{3.132}$$

$$\omega^2 = \frac{(2\bar{\Omega}\cdot\bar{k})^2}{k^2}$$

again with the high-frequency cut-off:

$$\omega_i = 2\Omega \frac{n}{\sqrt{l^2 + m^2 + n^2}} \leq 2\Omega \tag{3.133}$$

circular orbits in the plane of the wave front.

3.12 Stratified fluids, internal gravity waves

In a stably stratified fluid ($N > 0$) the motion resulting from the archimedean restoring force can overshoot the equilibrium position and give rise to *gravitational oscillations* (internal waves) having the Brunt-Väisälä frequency N.

For an incompressible fluid

$$N^2(r) = -g(r)\rho^{-1}(r)\frac{\mathrm{d}\rho(r)}{\mathrm{d}r} \quad (*) \tag{3.134}$$

By analogy with the Ekman number ($\nu/\Omega L^2$) we shall introduce here a stratification scaling factor defined as follows:

$$E_s = \frac{\nu}{NL^2} = \nu \Big/ \left(L\sqrt{\frac{\Delta\rho}{\rho} gL} \right) \tag{3.135}$$

According to eqn. 2.26, $(\Delta\rho/\rho)\,g = -g\alpha T$ and, by using the scaling factor $\left(L \Big/ \frac{\Delta\rho}{\rho} g \right)^{1/2}$ as done by Greenspan (1969, pp. 17–18), we can write linear dimensionless equations as follows:

$$\left.\begin{aligned} \nabla\cdot\bar{q} &= 0 \\ \frac{\partial}{\partial t}\bar{q} &= -\nabla p - s\hat{k} + E_s\nabla^2\bar{q} \\ \frac{\partial}{\partial t}T + \bar{q}\cdot\nabla T &= \frac{E_s}{P}\nabla^2 T \\ s &= -T \end{aligned}\right\} \tag{3.136}$$

The parameter s is needed to make the equations totally dimensionless. Therefore Greenspan introduces:

a stratification degree represented by $\Delta\rho$

the Froude number $\qquad F_R = \Omega^2 L/g$

a density ratio $\qquad И = \Delta\rho/\rho_0 \qquad$ (3.137)

the "internal" Froude number $\qquad f_r = F_R/И = \Omega^2 L \Big/ \frac{\Delta\rho}{\rho} g$

* See Crossley and Rochester (1980) concerning the physical meaning of N^2 in this case.

in this way the density perturbation is introduced by a dimensionless parameter.

$P = \nu/\chi$ is the Prandtl number.

Taking here $(U\ V\ W)$ for the *displacement*, one has

$$\bar{q} = \bar{q}(\dot{U}, \dot{V}, \dot{W}) \qquad \bar{q}\cdot\nabla T = \bar{q}\cdot\hat{k} = \dot{W}$$

and

$$\left.\begin{aligned}\left(\frac{\partial}{\partial t} - E_s\nabla^2\right)\dot{U} &= -\frac{\partial p}{\partial x}\\ \left(\frac{\partial}{\partial t} - E_s\nabla^2\right)\dot{V} &= -\frac{\partial p}{\partial y}\end{aligned}\right\}\left(\frac{\partial}{\partial t} - E_s\nabla^2\right)\frac{\partial \dot{W}}{\partial z} = \frac{\partial^2 p}{\partial x^2} + \frac{\partial^2 p}{\partial y^2} \tag{3.138}$$

$$\left(\frac{\partial}{\partial t} - E_s\nabla^2\right)\dot{W} = -\frac{\partial p}{\partial z} - T$$

Thus:
$$\left(\frac{\partial}{\partial t} - E_s\nabla^2\right)\frac{\partial \dot{W}}{\partial z} = -\frac{\partial^2 p}{\partial z^2} - \frac{\partial T}{\partial z} \tag{3.139}$$

and, from (3.138):

$$-\frac{\partial T}{\partial z} = \nabla^2 p \tag{3.140}$$

$$\left[\frac{\partial}{\partial t} - (E_s/P)\nabla^2\right]T = -\bar{q}\cdot\nabla T = -\bar{q}\cdot\hat{k} = -\dot{W} \tag{3.141}$$

Applying the operator

$$\left(\frac{\partial}{\partial t} - E_s\nabla^2\right)\frac{\partial}{\partial z}$$

to (3.141) we have:

$$\left(\frac{\partial}{\partial t} - \frac{E_s}{P}\nabla^2\right)\left(\frac{\partial}{\partial t} - E_s\nabla^2\right)\frac{\partial T}{\partial z} = -\left(\frac{\partial}{\partial t} - E_s\nabla^2\right)\frac{\partial \dot{W}}{\partial z} \tag{3.142}$$

or, using (3.138), (3.139) and (3.140), *a Greenspan equation for stratified fluids:*

$$\left(\frac{\partial}{\partial t} - \frac{E_s}{P}\nabla^2\right)\left(\frac{\partial}{\partial t} - E_s\nabla^2\right)\nabla^2 p + \left(\frac{\partial^2}{\partial x^2} + \frac{\partial^2}{\partial y^2}\right)p = 0 \tag{3.143}$$

If $P = 1$, that is $\nu \approx \chi$, one has:

$$\left(\frac{\partial}{\partial t} - E_s\nabla^2\right)^2\nabla^2 p + \left(\frac{\partial^2}{\partial x^2} + \frac{\partial^2}{\partial y^2}\right)p = 0 \tag{3.144}$$

A very important consequence of equation (3.144) *is that stratification has an effect similar to rotation.*

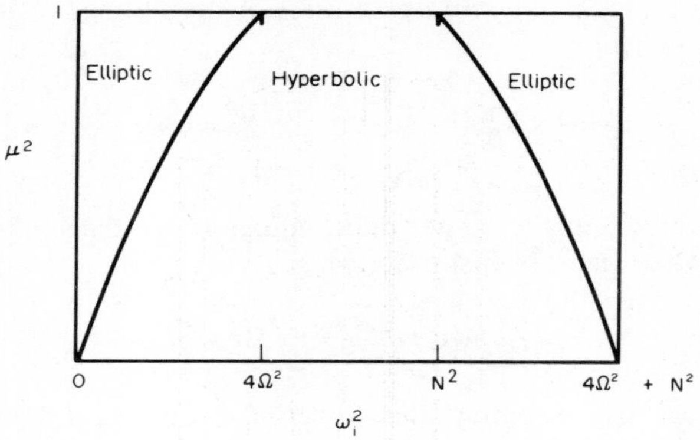

FIGURE 3.5 The regimes delineated by eqn (3.182). The drawing is scaled for $N^2 = 8\Omega^2$ in the interest of clarity. The tidal regime is $\omega_i^2 < 4\Omega^2$ in which the governing partial differential equation is mixed. After J. W. Miles, 1974.

In the case of stratification, a length (x or y) taken along a constant density surface plays the same role as a length taken along the rotation axis (z) in the case of rotation. It results that Taylor columns must have their counterpart in stratified media and that, in the laboratory, rotation can be used to simulate stratification (Veronis, 1970).

Let us consider an initial situation, at rest:

$$\vec{q}_0 \equiv 0 \qquad p_Z = -g\rho_0(z) \tag{3.145}$$

and let us introduce a *small* perturbation:

$$\begin{cases} \vec{q}^* = \vec{q}_0 + \vec{q} \\ \rho^* = \rho_0 + \rho \\ p^* = p_0 + p \end{cases} \tag{3.146}$$

The conservation laws for mass and momentum are respectively:

mass:
$$\frac{\partial \rho^*}{\partial t} + \nabla \cdot (\rho^* \vec{q}^*) = 0 \tag{3.147}$$

momentum:
$$\frac{D\vec{q}^*}{Dt} + f\hat{k} \wedge \vec{q}^* = -\frac{1}{\rho^*}\nabla p^* - g\hat{k} \tag{3.148}$$

Here f is the inertial frequency 2Ω. Meteorologists however use $f = 2\Omega \cos\theta$, the Coriolis parameter (see eqn. 3.72).

We have seen that for velocities much lower than seismic velocities we can consider the fluid as incompressible. Then

$$\frac{D\rho^*}{Dt} = 0 \qquad \nabla \cdot \vec{q}^* = 0 \qquad N^2 = -\frac{g}{\rho}\frac{\partial \rho}{\partial z} \tag{3.149}$$

so that we will obtain only *transverse waves.* By neglecting the advection we linearize the equations:

$$(\rho_0+\rho)\left(\frac{\partial \bar{q}_0}{\partial t}+\frac{\partial \bar{q}}{\partial t}+f\hat{k}\wedge\bar{q}_0+f\hat{k}\wedge\bar{q}\right)=-\nabla(p+p_0)-g\hat{k}(\rho_0+\rho) \tag{3.150}$$

Then, by subtracting the conservation equation for the initial state and keeping only the terms of first order we get:

$$\rho_0\frac{\partial \bar{q}}{\partial t}+\rho_0 f\hat{k}\wedge\bar{q}=-\nabla p-g\hat{k}\rho \tag{3.151}$$

Let us consider that the perturbation is periodic in $e^{-i\omega_i t}$, and $\bar{q}=\bar{q}(u,v,w)$, then:

$$\left.\begin{aligned}
-i\omega_i u-fv &= -\frac{1}{\rho_0}\frac{\partial p}{\partial x}\\
-i\omega_i v+fu &= -\frac{1}{\rho_0}\frac{\partial p}{\partial y}\\
-i\omega_i w &= -\frac{1}{\rho_0}\frac{\partial p}{\partial z}-g\frac{\rho}{\rho_0}
\end{aligned}\right\} \tag{3.152}$$

while

$$\frac{\partial \rho}{\partial t}+w\frac{\partial \rho}{\partial z}=0 \tag{3.153}$$

so that

$$-i\omega_i\rho+w\frac{\partial \rho}{\partial z}=0 \tag{3.154}$$

The third equation

$$-i\omega_i w\rho_0=-\frac{\partial p}{\partial z}-g\rho$$

or

$$\omega_i^2 w\rho_0=-i\omega_i\frac{\partial p}{\partial z}-i\omega_i g\rho$$

can be written, by introducing (3.154):

$$\omega_i^2 w\rho_0=-i\omega_i\frac{\partial p}{\partial z}-w\frac{\partial \rho}{\partial z}g \tag{3.155}$$

so that, with the definition (3.149), we obtain

$$\boxed{\rho_0(N^2-\omega_i^2)w=i\omega_i\frac{\partial p}{\partial z}} \tag{3.156}$$

Let us now solve the first two equations (3.152) in u and v:

$$\begin{cases} u = \dfrac{1}{\rho_0}\dfrac{-i\omega_i p_x + f p_y}{\omega_i^2 - f^2}, & p_x = \dfrac{\partial p}{\partial x} \\[2ex] v = \dfrac{1}{\rho_0}\dfrac{-i\omega_i p_y - f p_x}{\omega_i^2 - f^2}, & p_y = \dfrac{\partial p}{\partial y} \end{cases} \tag{3.157}$$

thus

$$\left\{ \begin{aligned} \frac{\partial u}{\partial x} &= \frac{1}{\rho_0(\omega_i^2 - f^2)}\left(-i\omega_i \frac{\partial^2 p}{\partial x^2} + f\frac{\partial^2 p}{\partial x\,\partial y}\right) \\ \frac{\partial v}{\partial y} &= \frac{1}{\rho_0(\omega_i^2 - f^2)}\left(-i\omega_i \frac{\partial^2 p}{\partial y^2} - f\frac{\partial^2 p}{\partial x\,\partial y}\right) \end{aligned} \right\} \tag{3.158}$$

But

$$\frac{\partial u}{\partial x} + \frac{\partial v}{\partial y} = -\frac{\partial w}{\partial z}$$

and we obtain a second equation for w:

$$\boxed{\rho_0(\omega_i^2 - f^2)\frac{\partial w}{\partial z} - i\omega_i \nabla_H^2 p = 0}\,^* \tag{3.159}$$

To eliminate the pressure let us write these two equations:

$$\begin{aligned} \rho_0(N^2 - \omega_i^2)\nabla_H^2 w &= i\omega_i \nabla_H^2 \frac{\partial p}{\partial z} \\ (\omega_i^2 - f^2)\frac{\partial}{\partial z}\left(\rho_0 \frac{\partial w}{\partial z}\right) &= i\omega_i \nabla_H^2 \frac{\partial p}{\partial z} \end{aligned} \tag{3.160}$$

and subtracting:

$$\rho_0(N^2 - \omega_i^2)\nabla_H^2 w - (\omega_i^2 - f^2)\frac{\partial}{\partial z}\left(\rho_0 \frac{\partial w}{\partial z}\right) = 0$$

or, better:

$$\frac{1}{\rho_0}\frac{\partial}{\partial z}\left(\rho_0 \frac{\partial w}{\partial z}\right) - \left(\frac{N^2 - \omega_i^2}{\omega_i^2 - f^2}\right)\nabla_H^2 w = 0 \tag{3.161}$$

Finally, with the Boussinesq approximation:

$$\boxed{\frac{\partial^2 w}{\partial z^2} - \frac{N^2 - \omega_i^2}{\omega_i^2 - f^2}\nabla_H^2 w = 0} \tag{3.162}$$

* ∇_H^2 is the horizontal Laplacian (see pages 89 and 101, eqn. 3.77).

This equation is hyperbolic and a wave form solution exists only if $N^2 > \omega_i^2 > f^2$ (strongly stable core) or $N^2 < \omega_i^2 < f^2$ (weakly stable core). If we put $N = 0$ we evidently find again the Poincaré equation:

$$\frac{\partial^2 w}{\partial z^2}\left(1 - \frac{\omega_i^2}{\omega_i^2 - f^2}\right) + \frac{\omega_i^2}{\omega_i^2 - f^2}\nabla^2 w = 0$$

or

$$\nabla^2 w - \frac{f^2}{\omega_i^2}\frac{\partial^2 w}{\partial z^2} = 0 \tag{3.163}$$

with $f = 2\Omega$.

Note that if $N^2 = f^2 = 4\Omega^2$ (3.162) becomes a Laplace equation $\nabla^2 w = 0$.

Condition at the free surface

We write $\dfrac{\mathrm{d}p}{\mathrm{d}t} = 0$ or $-i\omega_i p + w\dfrac{\partial p}{\partial z} = 0$ at $z = 0$ (3.164)

or

$$-i\omega_i p - w\rho_0 g = 0 \tag{3.165}$$

so that we can eliminate the pressure from eqn (3.159)

$$\boxed{(\omega_i^2 - f^2)\frac{\partial w}{\partial z} + g\nabla_H^2 w = 0 \quad \text{at } z = 0} \tag{3.166}$$

Condition at the solid surface

$$\boxed{w = 0 \quad \text{at } z = -D} \tag{3.167}$$

Let us seek for solutions in the form

$$w = \mathrm{e}^{(ilx - i\omega_i t)}\, w(z) \tag{3.168}$$

we get

$$\frac{\partial^2 w}{\partial z^2} - R^2\left(\frac{\partial^2 w}{\partial x^2} + \frac{\partial^2 w}{\partial y^2}\right) = 0 \tag{3.169}$$

where

$$R^2 = \frac{N^2 - \omega_i^2}{\omega_i^2 - f^2} > 0 \text{ by hypothesis} \tag{3.170}$$

this gives

$$\frac{\partial^2 w}{\partial z^2} + l^2 R^2 w = 0 \tag{3.171}$$

and if $\underline{R^2 > 0}$

$$w = \mathrm{e}^{(ilx - i\omega_i t)} \cdot \sin\{lR(z + D)\} \tag{3.172}$$

where $w = 0$ upon $z = -D$. If, moreover $w = 0$ upon $z = 0$

$$\sin(lRD) = 0 \tag{3.173}$$

and

$$l_n RD = \pm n\pi \tag{3.174}$$

Finally we obtain the solution

$$w = e^{(ilx - i\omega_i t)} \sin\left\{\pm \frac{n\pi}{D}(z + D)\right\} \tag{3.175}$$

$$u = iRwe^{(ilx - i\omega_i t)} \cos\left\{\pm \frac{n\pi}{D}(z + D)\right\} \tag{3.176}$$

and, using (3.162) we obtain:

$$\frac{N^2 - \omega_i^2}{\omega_i^2 - f^2} = \left(\frac{n\pi}{lD}\right)^2 \tag{3.177}$$

i.e

$$\omega_i^2 = \frac{N^2 + f^2 (n\pi/lD)^2}{1 + (n\pi/lD)^2} \tag{3.178}$$

For $\quad l \to \infty \quad \omega_i \to N \quad 2\Omega = f < \omega_i < N$

$\quad l \to 0 \quad \omega_i \to f \quad (R^2 > 0)$

the character of inertial waves ($\omega_i \leq 2\Omega$) is destroyed by stratification

$\omega_i = 2\Omega$ is a low-frequency cut-off.

The governing equation for pressure can be obtained by deriving eqn. 3.158:

$$\rho_0 (N^2 - \omega_i^2) \frac{\partial w}{\partial z} = i\omega_i \frac{\partial^2 p}{\partial z^2} \tag{3.179}$$

so that eqn (3.159) becomes

$$\frac{\omega_i^2 - f^2}{N^2 - \omega_i^2} \frac{\partial^2 p}{\partial z^2} - \nabla_H^2 p = 0 \tag{3.180}$$

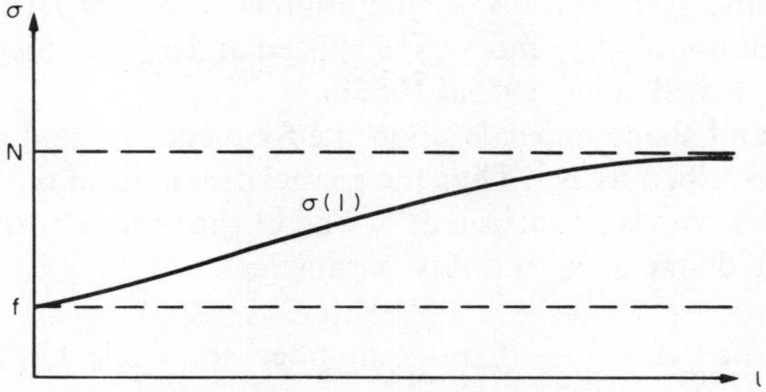

FIGURE 3.6

or

$$\boxed{\nabla^2 p + \frac{N^2 - f^2}{\omega_i^2 - N^2}\frac{\partial^2 p}{\partial z^2} = 0} \tag{3.181}$$

(Friedlander and Siegmann, 1982), a generalization of the Poincaré equation (3.97) (where $N^2 = 0$).

As in the previous case of purely inertial waves (§3.7), the governing equation for internal waves in a spherical rotating stratified fluid is of mixed type. It is hyperbolic when

$$D = \omega_i^2 - (N^2 + 4\Omega^2) + \left(\frac{N}{\omega_i}\right)^2 f^2 < 0 \tag{3.182}$$

$$f = 2\Omega \cos\theta \quad \text{(see eqn 3.72)}$$

and elliptic when $D > 0$ (Miles, 1974; Friedlander, 1982).

For each mode of frequency ω_i, the hyperbolic and elliptic regimes are delineated by a "turning (or critical) surface", which is a functional of N, f and ω_i.

The equation of this surface results from eqn (3.182):

$$N^2[4\Omega^2\cos^2\theta - \omega_i^2] + \omega_i^2(\omega_i^2 - 4\Omega^2) = 0 \tag{3.183}$$

from which cut-off points on the boundary sphere are given by

$$\cos^2\theta = \frac{\omega_i^2(N^2 + 4\Omega^2 - \omega_i^2)}{4\Omega^2 N^2}$$

If $0 \le \cos\theta \le 1$ — θ is real and critical surfaces exist inside the boundary sphere, the solution is not global and baroclinic modes are "trapped" between inertial latitudes (i.e. the so-called "equatorial trapping" of low frequency planetary waves);

if $\cos\theta > 1$ or $\cos\theta < 0$ — the critical surfaces are imaginary and one has global oscillations in the sphere.

The nature of the surface (3.183) is determined for each frequency by the spatial distribution of N^2 and $\cos^2\theta$ (Friedlander and Siegmann, 1982; Crossley, 1984; Friedlander, 1985a, 1985b).

Its position and shape depends upon the frequency ω_i and of the density stratification described by N^2. Thus the spatial distribution of internal waves inside the Earth's core is a function of ω_i and of the radial distribution of N^2.

One can thus define as a "stability parameter"

$$S = \frac{N}{2\,\Omega\cos\theta} \quad \text{or} \quad A = \frac{N^2}{4\Omega^2} \quad \text{(Froude number, see Table 1.7, p. 38)} \tag{3.184}$$

if $A > 1$ the system is strongly stable; if $A < 1$ it is weakly stable.

3.13 Turbulence in hydrodynamics

"It is not very likely that science will ever achieve a complete understanding of the mechanism of turbulence because of its extremely complicated nature" (Schlichting, 1968, p. 544)

What is turbulence?

Turbulence appears when the Reynolds number is much greater than unity, and exceeds a critical value depending on the conditions.

The elements in movement, exhibiting sudden and chaotic variations in space and time, have lost their memory of previous situations. Turbulence defeats determinism so that predictability is no more possible; one cannot any more treat the velocity $\bar{q}$ as a *know* function of position and time $\bar{q}(\bar{x}, t)$ as in the case of laminar flow.

The result is a loss of small-scale correlation between successive situations. Therefore only statistical methods can be applied to try to describe the phenomena.

In the turbulent motion, $\bar{q}$ thus contains a random part of which only "mean" statistical properties can be considered, while their detailed properties are far too complicated to be described analytically.

This random movement can be either a random velocity field or a random superposition of interacting waveform movements. To treat this problem one has to decide to separate, *by convention*, the velocity field into two parts:

$\langle \bar{q} \rangle$ a large-scale flow;
$\bar{q}'$ a small-scale turbulence.

However there exists no logical criterion on how to operate this separation. One puts:

$$\bar{q} = \langle \bar{q} \rangle + \bar{q}' \quad \text{with} \quad \langle \bar{q}' \rangle = 0$$

and

$$\langle \langle \bar{q} \rangle + \bar{q}' \rangle = \langle \bar{q} \rangle \tag{3.185}$$

This averaging can be conducted into different ways:

1. One can admit that the *length scale* of random movements, or the wave length of random waves l_0 is much smaller than L, the linear dimension of the container. Then, at intermediary scale:

$$l_0 \ll a \ll L \tag{3.186}$$

one can define an average

$$\langle \psi(\bar{x}, t) \rangle_a = \frac{3}{4\pi a^3} \int_{|\xi| < a} \psi(\bar{x} + \bar{\xi}, t)\, d^3\xi \tag{3.187}$$

which is insensitive to the exact value of a, provided (3.186) is satisfied.

2. One could use the *time scale* instead of the spatial scale and consider a scale of global variation T and a scale characteristic of the fluctuations t_0. Then, for τ such that

$$t_0 \ll \tau \ll T \tag{3.188}$$

one can define an average

$$\langle \psi(\bar{x}, t) \rangle_\tau = \frac{1}{2\tau} \int_{-\tau}^{+\tau} \psi(\bar{x}, t + \tau')\, d\tau' \tag{3.189}$$

which is insensitive to the exact value of τ provided (3.188) is satisfied.

3. One can even define the mean field in terms of averaging over an *azimuthal angle* λ:

$$\langle \psi(r, z, \lambda) \rangle_\lambda = \frac{1}{2\pi} \int_0^{2\pi} \psi(r, z, \lambda')\, d\lambda' \tag{3.190}$$

The ergodic hypothesis states that (3.187) is equivalent to (3.189) or (3.190)

Reynolds stress tensor

With these premises, let us rewrite the first Navier equation with the advection term:

$$\left[\frac{\partial}{\partial t} + (\langle u \rangle + u')\frac{\partial}{\partial x} + (\langle v \rangle + v')\frac{\partial}{\partial y} + (\langle w \rangle + w')\frac{\partial}{\partial z}\right](\langle u \rangle + u')$$

$$-2\Omega(\langle v \rangle + v') = -\frac{1}{\rho}\frac{\partial \langle p \rangle}{\partial x} - \frac{1}{\rho}\frac{\partial p'}{\partial x} + \nu \nabla^2 u \tag{3.191}$$

Let us also develop

$$\frac{\partial}{\partial x}\langle u'u' \rangle + \frac{\partial}{\partial y}\langle v'u' \rangle + \frac{\partial}{\partial z}\langle w'u' \rangle = u'\frac{\partial u'}{\partial x} + v'\frac{\partial u'}{\partial y} + w'\frac{\partial u'}{\partial z}$$

$$+ u'\left[\frac{\partial u'}{\partial x} + \frac{\partial v'}{\partial y} + \frac{\partial w'}{\partial z}\right] \tag{3.192}$$

where the last term is zero.

Averaging upon this equation then gives:

$$\frac{\partial \langle u \rangle}{\partial t} + \langle u \rangle \frac{\partial \langle u \rangle}{\partial x} + \langle v \rangle \frac{\partial \langle u \rangle}{\partial y} + \langle w \rangle \frac{\partial \langle u \rangle}{\partial z} - 2\Omega \langle v \rangle =$$

$$= -\frac{1}{\rho}\frac{\partial \langle p \rangle}{\partial x} + \nu \nabla^2 \langle u \rangle - \frac{\partial}{\partial x}\langle u'u' \rangle - \frac{\partial}{\partial y}\langle v'u' \rangle - \frac{\partial}{\partial z}\langle w'u' \rangle \tag{3.193}$$

which shows that the equation relative to large-scale flow velocities (u, v, w) contains three quadratic terms in the fluctuations which do not disappear when averaging.

As an example, $\rho\langle v'u'\rangle$ represents a mean flow of moment $\rho u'$ in the x component across a surface $y = c^{\text{ste}}$. Let us put

$$\langle v'u'\rangle = -\frac{\tau_{xy}}{\rho} \tag{3.194}$$

the dimensions of the left side are $(LT^{-1})^2$ while in the right side we have:

$$\frac{\text{Force}}{\text{Surface}} \times \frac{\text{Volume}}{\text{Mass}} = L^2T^{-2}.$$

as this flow is equivalent to a stress τ exerted upon the large-scale flow. The Navier equations then take the form:

$$\frac{\partial u}{\partial t} + u\frac{\partial u}{\partial x} + v\frac{\partial u}{\partial y} + w\frac{\partial u}{\partial z} - 2\Omega v$$

$$= -\frac{1}{\rho}\frac{\partial p}{\partial x} + \frac{1}{\rho}\left\{\frac{\partial \tau_{xx}}{\partial x} + \frac{\partial \tau_{xy}}{\partial y} + \frac{\partial \tau_{xz}}{\partial z}\right\} + \nu\nabla^2 u \tag{3.195}$$

where

$$\begin{aligned} \tau_{xx} &= -\rho\langle u'u'\rangle, \ldots \\ \tau_{xy} = \tau_{yx} &= -\rho\langle u'v'\rangle, \ldots \end{aligned} \tag{3.196}$$

are called *Reynolds stresses.*

One can suppose—but this is quite arbitrary, over-simple and even perhaps doubtful—that such stresses act as they would at molecular scale, and that one can introduce linear relations of the same kind:

$$\frac{\tau_{xx}}{\rho} = 2A_H\frac{\partial u}{\partial x}, \quad \frac{\tau_{yy}}{\rho} = 2A_H\frac{\partial v}{\partial y}, \quad \frac{\tau_{zz}}{\rho} = 2A_v\frac{\partial w}{\partial z} \tag{3.197}$$

$$\left.\begin{aligned} \frac{\tau_{xy}}{\rho} &= \frac{\tau_{yx}}{\rho} = A_H\left(\frac{\partial v}{\partial x} + \frac{\partial u}{\partial y}\right) \\ \frac{\tau_{xz}}{\rho} &= \frac{\tau_{zx}}{\rho} = A_v\frac{\partial u}{\partial z} + A_H\frac{\partial w}{\partial x} \\ \frac{\tau_{yz}}{\rho} &= \frac{\tau_{zy}}{\rho} = A_v\frac{\partial v}{\partial z} + A_H\frac{\partial w}{\partial y} \end{aligned}\right\} \tag{3.198}$$

One introduces here two different coefficients, A_v and A_H, because there is no reason to believe that the *coefficients of turbulent viscosity* are the same in the vertical and in the horizontal components. They are properties of the motion, not of the matter like kinematic or dynamic viscosity, and there is no way to calculate them *a priori.**

* Turbulent viscosity is an empirical concept, hard to justify, even harder to quantify and impossible to derive rigorously (Pedlosky, 1979, p. 46).

Finally the turbulent Navier equations take the form:

$$\frac{\partial u}{\partial t}+u\frac{\partial u}{\partial x}+v\frac{\partial u}{\partial y}+w\frac{\partial u}{\partial z}-2\Omega v$$

$$=-\frac{1}{\rho}\frac{\partial p}{\partial x}+A_H\left(\frac{\partial^2 u}{\partial x^2}+\frac{\partial^2 u}{\partial y^2}\right)+A_v\frac{\partial^2 u}{\partial z^2}+\nu\nabla^2 u \quad (3.199)$$

Now, if one accepts the idea that $A_v = A_H = A$, these equations would be exactly similar to the laminar Navier equations but with a viscosity $\nu + A$, much more efficient because A can exceed ν by several orders of magnitude.

Oceanographers often consider, for the ocean:

$$A_v \sim 1 \text{ to } 10^3 \text{ stokes}$$
$$A_H \sim 10^5 \text{ to } 10^8 \text{ stokes}$$

while

$$\nu \sim 10^{-2} \text{ stokes}$$

In the Prandtl theory A is related to a transverse "mixing length" (more exactly to its square).

Kolmogorov theory

The typical behaviour of turbulent motion was described by Richardson (1922), who showed that turbulence results in a *hierarchy of eddies* of successive sizes or orders: the lack of stability of a large eddy stimulates a (non-linear) transfer of its energy to smaller eddies which, in their turn, destabilize and generate even smaller eddies. In this "*cascade*" process towards movements of smaller and smaller scale, the kinetic energy is transferred to this very small scale at which dissipation by viscosity occurs.

The theory of Kolmogorov (1941) is based upon such concepts. The main characteristics of turbulence are:

1. its *non-linearity*;
2. its *tridimensionality*;
3. the *transfer of energy to smaller and smaller spatial scales.*

Obviously the turbulent flow is a non-linear mechanical system with a very large number of degrees of freedom. It can be studied only by statistical means.

The size of the eddies can be characterized by a wave number k (modulus of a wave vector $\vec{k}$) so that the differential operators correspond to ik_j for $\partial/\partial x_j$ and $-k^2$ for Δ, and the energy dissipation due to molecular viscosity has a rate $-2\nu k^2 E(k)$ where $E(k)$ is the energy contained in all modes of modulus k.

In the Kolmogorov theory, energy is initially available or continuously restored by external forces, at large scale, that is around the *energetic wave number* k_e. This energy decreases as a result of:

1. a molecular viscous decrease which augments with the square of k, that is very considerably at small scales (k high) but negligibly at large scales (k small);
2. a non-linear *conservative* transfer towards smaller and smaller scales (higher wave numbers) due to the non-linear terms of advection; this is the "*inertial cascade*" towards very small scales where there is this "energy sink" corresponding to the molecular viscous dissipation.

As a matter of fact the dissipation becomes important for wave numbers higher than the "*Kolmogorov dissipation wave number* k_d". The size of the structures where viscous dissipation reaches the same order of magnitude as the inertial phenomena is of the order of $1/k_d$. There is thus an energy flow between k_e and k_d.

The theorem of Kolmogorov is expressed by the existence of:

1. An *inertial zone* corresponding to k values such that

$$k_e < k < k_d = \left(\frac{\varepsilon}{\rho v^3}\right)^{1/4} \tag{3.200}$$

where the energy does not depend on v:

$$E(k) = c_k \rho^{1/3} \varepsilon^{2/3} k^{-5/3} \tag{3.201}$$

c_k *is the Kolmogorov constant* ($c_k \sim 1.5$), ε being the rate of inertial cascade.

2. A *dissipative zone* corresponding to

$$k \gtrsim k_d \tag{3.202}$$

where the energy depends on v:

$$E(k) = f(k/k_d) \rho^{1/3} \varepsilon^{2/3} k^{-5/3} \tag{3.203}$$

One can establish that the Reynolds number is related to k_d:

$$\mathrm{Re} \sim \left(\frac{k_d}{k_e}\right)^{4/3} \tag{3.204}$$

so that the Kolmogorov law allows quantification of the existence of the inertial zone as a function of the Reynolds number. For very large Reynolds numbers this inertial zone develops considerably, but the dissipation rate at infinitesimal viscosity is not zero because v is then multiplied by an infinitely great dissipative wave number.

The existence of a non-zero dissipation at infinitesimal viscosity implies that the curl of the velocity becomes infinite, which means that zones of quasi-discontinuities appear in the flow (Obukhov model).

Remarks

It is not only the diffusion of momentum (Navier equation) which can be subject to turbulence, but each transport property can exhibit such behaviour:

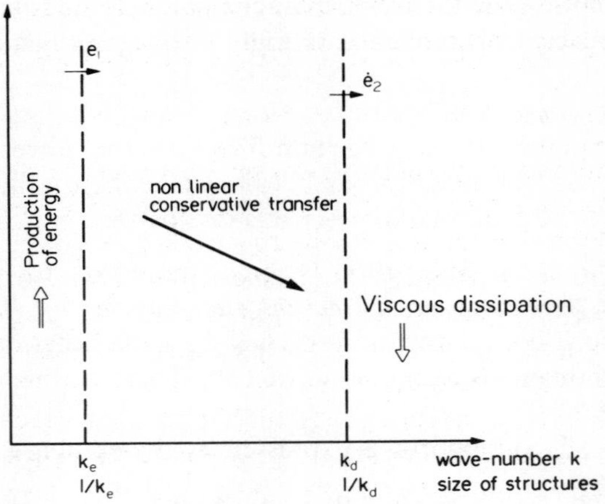

FIGURE 3.7 Kolmogorov theory.

concentration, heat, magnetic field. In a general way the transport equation

$$\frac{\partial S}{\partial t} + q_i \frac{\partial S}{\partial x_i} = \frac{\partial}{\partial x_i}\left(\beta_{ij}\frac{\partial S}{\partial x_j}\right) = \beta \frac{\partial^2 S}{\partial x_i \partial x_i} = \beta \nabla^2 S \tag{3.205}$$

(the last two terms in a case of isotropic homogeneous diffusion) will contain terms of turbulence

$$\langle q' s' \rangle = -D\frac{\partial S}{\partial x_i} = -D\left(\frac{\partial S_i}{\partial x_j} + \frac{\partial S_j}{\partial x_i}\right) \tag{3.206}$$

and if one puts

$$D + \beta = \beta' \tag{3.207}$$

the fundamental equation will keep the same form in all cases.

Backus (1968) argued that to raise ν to $10^8\,\mathrm{cm^2\,s^{-1}}$ with relative velocities less than $0.5\,\mathrm{cm\,s^{-1}}$, eddies 4000 km across would be necessary.

The geomagnetic maps look like meteorological maps, which leads one to assume that the core will be turbulent. Moreover the Reynolds number is very high ($R = E^{-1} = 10^{15}$), and this adds to the same feeling.

"If turbulence does occur to a serious extent, it is our view that the problems raised in the study of core motions are to a large extent insurmountable in the present state of knowledge" (Hide and Stewartson, 1972).

3.14 The boundary layers in hydrodynamics

The theory of boundary layers was introduced by Prandtl in 1904 under the name "*übergangschicht*" (transition layer) to eliminate the problem of discontinuities by inserting a thin layer where the transverse displacement varies very rapidly while the other characteristic quantities (such as pressure and radial displacement) remain practically constant in the layer.

Boundary conditions

The rheological property called viscosity originates in the cohesion and interaction between molecules: it is opposed to the transport of momentum (mv) and is responsible for the adherence of fluids at the surface of solids.

Viscosity controls the transfer of movement by a slipping between one sheet and the adjacent one: this involves a shear stress (τ) so that energy must be furnished to maintain the flow. This energy is dissipated into heat by exciting the oscillations of the molecules.

One then has to raise the question: what happens at a solid–liquid interface or "boundary"? Such a "boundary" is a discontinuity where the physical properties change abruptly, and consequently also the field variables. It results that the equations of mechanics cannot be solved in such a region. They must therefore be replaced by an *equal* number of discontinuity conditions across the boundary: each equation corresponds to a number of conditions, this number being equal to the order of the differential equation.

Thus, for a viscous fluid, as the Navier differential equation is one order higher than for a perfect fluid, one will have to add one more boundary condition. For the perfect fluid the usual logic condition is that the fluid cannot penetrate the solid, thus that the normal component of speed is zero at the boundary while the fluid can slip along the boundary:

$$(\bar{q} \cdot \hat{n}) = 0 \tag{3.208}$$

Now if we accepted such a discontinuity in the tangential velocity for a viscous fluid, an infinite shear stress would be implied by the infinite gradient velocity. This cannot be, so that the additional condition will be a non-slip condition, thus

$$(\bar{q} \wedge \hat{n}) = 0 \tag{3.209}$$

It is well known, on *experimental grounds*, that adherence to the boundary always takes place even if the viscosity is extremely small. There is *always* a thin sheet inside which the velocity decreases up to zero on the wall.

Thus, the viscosity ν being as small as we want, there are always *two boundary conditions*, while for a perfect fluid $\nu = 0$, there is only one boundary condition involving the normal component alone.

Therefore a perfect fluid motion with $\nu = 0$ *is in no way* an approximation for the viscous fluid motion. The effect of viscosity is indeed to prevent *any* discontinuity in transverse displacements along interfaces. The interface is a source of vorticity which diffuses away by viscous action.

This annulation of the tangential velocity is reached inside a thin shear layer close to the boundary which is characterized by a strong transverse gradient of the velocity. It is therefore the place where dissipation occurs.

This *boundary layer* can be defined as the sheet, of thickness δ, where the viscous terms $\nu\nabla^2\bar{q}$ (dimension $(\nu U)/\delta^2$) become of the same order of magnitude as the non-linear advection terms $\bar{q}\cdot\nabla\bar{q}$ (dimension U^2/L, with L being a characteristic length), that is:

$$\frac{U^2}{L} \sim \nu\frac{U}{\delta^2} \tag{3.210}$$

Using the *Reynolds number*

$$\mathrm{R} = \frac{UL}{\nu} \tag{3.211}$$

we obtain

$$\delta \sim \left(\frac{\nu L}{U}\right)^{1/2} = \frac{L}{R^{1/2}} \tag{3.212}$$

If

$$U = \Omega L \tag{3.213}$$

one has

$$\delta \sim \sqrt{\frac{\nu}{\Omega}} \tag{3.214}$$

However this thickness, clearly proportional to $\nu^{1/2}$, cannot be defined rigorously because the influence of viscosity decreases asymptotically outwards. It can be estimated according to one of the following rules:

1. distance where $V = 0.99\, V_\infty$
2. intersection of the tangent at the origin with the asymptote
3. area equality:

$$\delta = \frac{1}{V_\infty}\int_0^\infty (V_\infty - V)dn \text{ (dashed area on figure 3.8)} \tag{3.215}$$

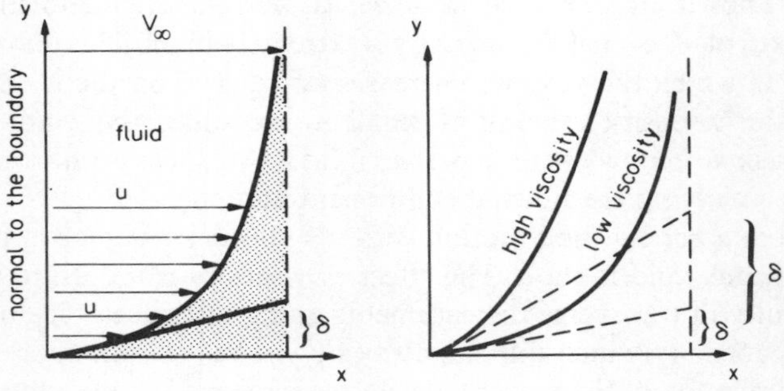

FIGURE 3.8 Boundary layers for different viscosities but $\dfrac{\partial v}{\partial x} \ll \dfrac{\partial u}{\partial y}$.

(we call V_∞ the velocity very far from the boundary). If the viscosity is high the gradient of the velocity (for a given stress) will be small and the layer very thick.

It is not difficult to show that the velocity profile is parabolic (figure 3.8).

Boundary layer equations

Even linearized, the Navier–Stokes equations are still too complicated. The procedure used as the "standard boundary layer theory" rests on the postulate that one can split the problem into two parts; a postulate which is justified *a posteriori* from the obtained results. The liquid body is accordingly divided into two regions where different systems of equations are to be solved:

1. In the bulk of the body the viscosity plays a very minor role. One drops the term $\nu\nabla^2\bar{q}$ and the equations can be integrated. However the solution obtained in this way cannot fit the conditions of non-penetration and non-slip at the boundary.
2. In a thin boundary layer, sticking on the solid–liquid interface we must keep the term $\nu\nabla^2\bar{q}$ as the viscosity multiplies the gradient of the velocity normal to the interface which is very strong. However we can simplify the equations by neglecting the other gradients

$$\left(\frac{\partial}{r\partial\theta}, \frac{\partial}{r\sin\theta\,\partial\lambda}\right)$$

   with respect to this normal gradient $(\partial/\partial r)$ and this is the characteristic feature in a boundary layer.

The role of the boundary layer is to match the conditions of the flow in the bulk of the body on its internal side and on its external side to match the non-slipping conditions on the solid boundary of the container. This shows that the boundary layer acts essentially for the *transmission of information* from the solid boundary to the bulk of the liquid body.

Inside this boundary layer the intensity of the gradient of velocity is inversely proportional to the thickness of the layer. This of course essentially depends upon the viscosity (laminar or turbulent).

In the typical case of the Earth's liquid core, any impulse changing the rotation parameters of the mantle (such as nutations or variations of speed) will have as a consequence the formation of an Ekman boundary layer transmitting the new rotation parameters and a viscous diffusion of vorticity to the liquid body to adjust it to a new steady rotation (transient motions, spin-up, spin-down and spin-over effects are associated with this new situation).

Let us consider the case of a *bidimensional incompressible fluid*. In the bulk of the fluid body, we drop the viscous dissipation term from the equations and

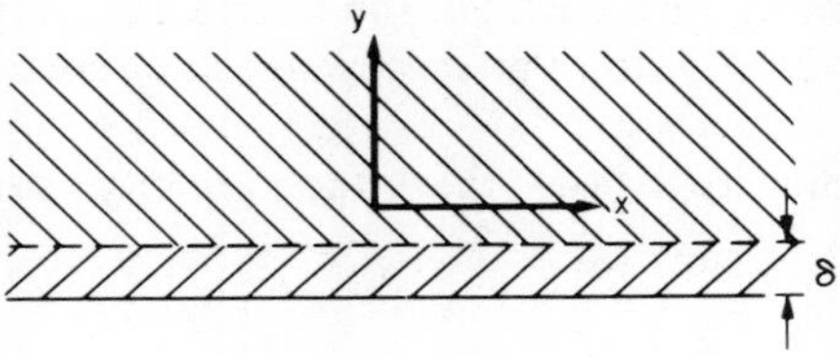

FIGURE 3.9 Boundary layer.

write

$$\left.\begin{aligned}\frac{\partial u}{\partial t}+u\frac{\partial u}{\partial x}+v\frac{\partial u}{\partial y}&=-\frac{1}{\rho}\frac{\partial p}{\partial x}\\ \frac{\partial v}{\partial t}+u\frac{\partial v}{\partial x}+v\frac{\partial v}{\partial y}&=-\frac{1}{\rho}\frac{\partial p}{\partial y}\\ \frac{\partial u}{\partial x}+\frac{\partial v}{\partial y}&=0\end{aligned}\right\}\tag{3.216}$$

In the boundary layer this term plays a major role but we operate another simplification: $\partial u/\partial y$ is very large inside this layer which makes $\nu(\partial^2 u/\partial y^2)$ non-negligible at all, while $\nu(\partial^2 u/\partial x^2)$ can be neglected. The boundary layer equations are thus:

$$\left.\begin{aligned}\frac{\partial u}{\partial t}+u\frac{\partial u}{\partial x}+v\frac{\partial u}{\partial y}&=-\frac{1}{\rho}\frac{\partial p}{\partial x}+\nu\frac{\partial^2 u}{\partial y^2}\\ \frac{\partial v}{\partial t}+u\frac{\partial v}{\partial x}+v\frac{\partial v}{\partial y}&=-\frac{1}{\rho}\frac{\partial p}{\partial y}+\nu\frac{\partial^2 v}{\partial y^2}\\ \frac{\partial u}{\partial x}+\frac{\partial v}{\partial y}&=0\end{aligned}\right\}\tag{3.217}$$

with the boundary conditions: at $y=0$: $u=v=0$

at $y=\infty$: $u=U(x,t)$

The bidimensional case allows another simplification by the introduction of a stream function $\psi(x,y,t)$ defined by

$$u=\frac{\partial\psi}{\partial y}\qquad v=-\frac{\partial\psi}{\partial x}\tag{3.218}$$

to satisfy the incompressibility condition. Then the boundary layer equation becomes:

$$\frac{\partial^2\psi}{\partial y\partial t}+\frac{\partial\psi}{\partial y}\frac{\partial^2\psi}{\partial x\partial y}-\frac{\partial\psi}{\partial x}\frac{\partial^2\psi}{\partial y^2}=-\frac{1}{\rho}\frac{\partial p}{\partial x}+\nu\frac{\partial^3\psi}{\partial y^3}\tag{3.219}$$

at the wall: $\partial\psi/\partial y=\partial\psi/\partial x=0$
at $t=0$ the velocity distribution is prescribed.

In the first equation (3.217) (equation in $\partial u/\partial t$) all terms are of order 1 and thus so is $\partial p/\partial x$, but in the second equation (equation in $\partial v/\partial t$) all terms (but $\partial p/\partial y$) are of order δ. *Thus* $\partial p/\partial y$ must be of order δ. The pressure increase across the boundary layer is thus of the order δ^2, that is negligible.

Consequently the pressure in a direction normal to the boundary layer is practically constant, and can be assumed equal to that at its outer edge where its value is determined by the outer flow (Slichting, 1968, p. 120).

Transformation of the boundary layer equations into the heat conduction equation (von Mises, 1927)

Still in the case of a *two-dimensional incompressible* flow, von Mises introduces as new variables:

$$\xi = x \qquad \eta = \psi \tag{3.220}$$

so that one can write

$$\left.\begin{aligned} \frac{\partial u}{\partial x} &= \frac{\partial u}{\partial \xi}\frac{\partial \xi}{\partial x} + \frac{\partial u}{\partial \eta}\frac{\partial \eta}{\partial x} = \frac{\partial u}{\partial \xi} - v\frac{\partial u}{\partial \psi} \\ \frac{\partial u}{\partial y} &= \frac{\partial u}{\partial \xi}\frac{\partial \xi}{\partial y} + \frac{\partial u}{\partial \eta}\frac{\partial \eta}{\partial y} = 0 + u\frac{\partial u}{\partial \psi} \end{aligned}\right\} \tag{3.221}$$

Then, for a *steady* flow ($\partial u/\partial t = 0$), the boundary layer equation (3.217) becomes

$$u\frac{\partial u}{\partial \xi} + \frac{1}{\rho}\frac{\partial p}{\partial \xi} = \nu u \frac{\partial}{\partial \psi}\left(u\frac{\partial u}{\partial \psi}\right) \tag{3.222}$$

$$\left(\text{one has indeed } \frac{\partial}{\partial y}\left(u\frac{\partial u}{\partial \psi}\right) = \frac{\partial}{\partial \eta}\left(u\frac{\partial u}{\partial \psi}\right)\frac{\partial \eta}{\partial y} \text{ with } \frac{\partial \eta}{\partial y} = \frac{\partial \psi}{\partial y} = u\right)$$

If we now put, as the reduced pressure:

$$G = p + \frac{1}{2}\rho u^2 \tag{3.223}$$

(cf. also Bernoulli's equation), the equation can be written

$$\frac{\partial G}{\partial \xi} = \nu u \frac{\partial^2 G}{\partial \psi^2} \tag{3.224}$$

which is similar to the heat equation

$$\frac{\partial T}{\partial t} = \chi \frac{\partial^2 T}{\partial x^2} \tag{3.225}$$

both are parabolic differential equations but (3.224) is non-linear because of νu, u depending on x and G.

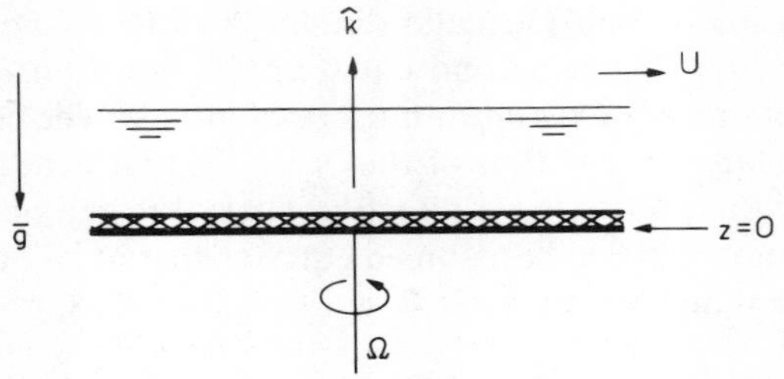

FIGURE 3.10.

3.15 Rotating fluid—Elementary properties of the viscous Ekman boundary layer

We will consider here a fluid rotating with the angular velocity Ω as shown on figure 3.10 as, for the existence of an Ekman boundary layer, there must be a component of $\bar{\Omega}$ orthogonal to the boundary. The Navier equations are

$$\begin{cases} \dfrac{\partial u}{\partial t} + u\dfrac{\partial u}{\partial x} + v\dfrac{\partial u}{\partial y} + w\dfrac{\partial u}{\partial z} - fv = -\dfrac{1}{\rho}\dfrac{\partial p}{\partial x} + A_V\dfrac{\partial^2 u}{\partial z^2} + A_H\left(\dfrac{\partial^2 u}{\partial x^2} + \dfrac{\partial^2 u}{\partial y^2}\right) \\ \dots \\ \dots \\ f = 2\Omega \quad \text{Coriolis parameter} \end{cases} \tag{3.226}$$

with, as boundary conditions:

for $z \to \infty$: $u = U \quad v = 0 \quad w = 0$
for $z = 0$: $u = 0 \quad v = 0 \quad w = 0$

Far from the boundary we can consider the flow as geostrophic ($2\hat{k} \wedge \bar{q} = -\nabla\phi$). The geostrophic equilibrium (equilibrium between Coriolis force and pressure forces) would give, in the bulk of the liquid:

$$u = U = -\frac{1}{\rho f}\frac{\partial p}{\partial y} \quad v = 0 \quad w = 0 \tag{3.227}$$

but this equilibrium is, of course, destroyed in the vicinity of the boundary. It seems logical to propose a solution

$$u = u(z) \quad v = v(z) \quad w = w(z) \tag{3.228}$$

so that the equation of conservation of mass immediately gives

$$\frac{\partial w}{\partial z} = 0 \tag{3.229}$$

As the $\partial/\partial z$ gradient greatly exceeds the gradients $\partial/\partial x$ or $\partial/\partial y$ we have, everywhere:

$$w = 0 \quad \text{because } w = 0 \text{ on } z = 0 \tag{3.230}$$

The system of Navier equations then becomes:

$$\left.\begin{aligned} -fv &= -\frac{1}{\rho}\frac{\partial p}{\partial x} + A_V\frac{\partial^2 u}{\partial z^2} \\ fu &= -\frac{1}{\rho}\frac{\partial p}{\partial y} + A_V\frac{\partial^2 v}{\partial z^2} \\ g &= -\frac{1}{\rho}\frac{\partial p}{\partial z} \end{aligned}\right\} \tag{3.231}$$

it expresses the balance between Coriolis force, pressure gradient and viscous forces, the last one being nothing else than the hydrostatic equilibrium equation. As g is independent from x and y and with $\rho = c^{\text{ste}}$ one has

$$\frac{\partial}{\partial x}\left(\frac{\partial p}{\partial z}\right) = \frac{\partial}{\partial y}\left(\frac{\partial p}{\partial z}\right) = 0 \tag{3.232}$$

or

$$\frac{\partial}{\partial z}\left(\frac{\partial p}{\partial x}\right) = \frac{\partial}{\partial z}\left(\frac{\partial p}{\partial y}\right) = 0 \tag{3.233}$$

which shows that the horizontal gradient of pressure must be independent of z. For very great values of $z\,(z \to \infty)$

$$-\frac{1}{\rho}\frac{\partial p}{\partial x} = 0 \qquad fU = -\frac{1}{\rho}\frac{\partial p}{\partial y} \tag{3.234}$$

which must be true for any value of z because the gradients of p are independent of z. Then the perturbations of u and v in the vicinity of the boundary satisfy the relations:

$$\left.\begin{aligned} f\tilde{u} &= A_V\frac{\partial^2 \tilde{v}}{\partial z^2} \qquad & \tilde{u} &= u - U \\ -f\tilde{v} &= A_V\frac{\partial^2 \tilde{u}}{\partial z^2} \qquad & \tilde{v} &= v \end{aligned}\right\} \tag{3.235}$$

where $(\tilde{u}, \tilde{v})$ represents the discrepancy, with respect to the geostrophic equilibrium, due to the boundary. Eliminating $\tilde{v}$ gives

$$\frac{\partial^4 \tilde{u}}{\partial z^4} + \frac{f^2}{A_V^2}\quad \tilde{u} = 0 \tag{3.236}$$

whose general solution is

$$\tilde{u} = C_1 e^{(1+i)z/\delta} + C_2 e^{(1-i)z/\delta} + C_3 e^{-(1+i)z/\delta} + C_4 e^{-(1-i)z/\delta} \tag{3.237}$$

where

$$\delta = \left(\frac{A_V}{f/2}\right)^{1/2} \tag{3.238}$$

If there was no turbulence this should simply reduce to:

$$\delta = \left(\frac{\nu}{\Omega}\right)^{1/2} = LE^{1/2}$$

which has the dimension of a length: it is, as a matter of fact, the *thickness* of the Ekman boundary layer, zone where there is a balance between viscous and Coriolis force. It is *proportional to the square root of the kinematic viscosity* but inversely proportional to the square root of the angular velocity. For the first two solutions, the amplitudes exponentially increase when $z \to \infty$ and must, therefore be discarded by putting $C_1 = C_2 = 0$. On the other hand, as $\tilde{u} = -U$ on $z = 0$ (non-slip condition) one obtains:

$$C_3 = C_4 = -\frac{U}{2}$$

The solution is thus:

$$\left.\begin{aligned} \tilde{u} &= -U e^{-z/\delta} \cos(z/\delta) \\ \tilde{v} &= U e^{-z/\delta} \sin(z/\delta) \end{aligned}\right\} \tag{3.239}$$

and, consequently

$$u = U(1 - e^{-z/\delta} \cos(z/\delta)) \tag{3.240}$$

It is very convenient to "scale" the distances inside the boundary layer with the "*Ekman depth*" by introducing there a "*stretched coordinate*"

$$\xi = z/\delta = E^{-1/2} z \tag{3.241}$$

which varies from 0 to 1 inside the layer so that the distances are measured in units of the Ekman depth δ. This procedure has the advantage of suppressing the rapid variations of u and v inside this layer. Then solutions for a non-zero Rossby number can be obtained by expansions in power series of ε.

Let us write, with Ekman, the solution as

$$\begin{aligned} \tilde{u} &= U e^{-z/\delta} \cos(z/\delta + c) \\ \tilde{v} &= -U e^{-z/\delta} \sin(z/\delta + c) \end{aligned} \tag{3.242}$$

from which we derive:

$$\left\{\begin{aligned} \frac{\partial \tilde{u}}{\partial z} &= -\frac{\sqrt{2}}{\delta} U e^{-z/\delta} \sin(z/\delta + c + 45^\circ) \\ \frac{\partial \tilde{v}}{\partial z} &= -\frac{\sqrt{2}}{\delta} U e^{-z/\delta} \cos(z/\delta + c + 45^\circ) \end{aligned}\right\} \tag{3.243}$$

Admitting that a tangential stress has been applied along the axis Oy, one should have:

$$\rho \nu \left(\frac{\partial \tilde{u}}{\partial z} \right)_{z=0} = 0 \tag{3.244}$$

$$-\rho \nu \left(\frac{\partial \tilde{v}}{\partial z} \right)_{z=0} = \tau_{yz} = \tau \text{ (Reynolds stress)} \tag{3.245}$$

From the first condition one obtains $c = -45^\circ$ and from the second:

$$U = \frac{\tau \delta}{\sqrt{2}\,\rho \nu} \quad \text{or} \quad \tau = \sqrt{2}\,\frac{\rho \nu U}{\delta} \tag{3.246}$$

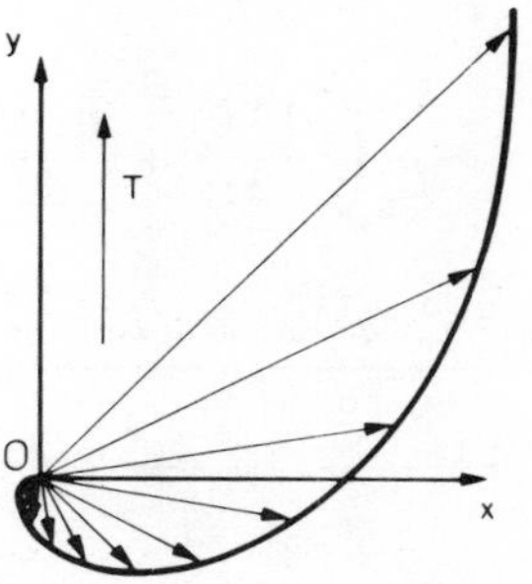

FIGURE 3.11 The original figure of the Ekman spiral published for the first time in Ekman (1905). This investigation was developed following a suggestion of Fridtjof Nansen to explain his observations made during the drift of his ship *Fram* in the Arctic.

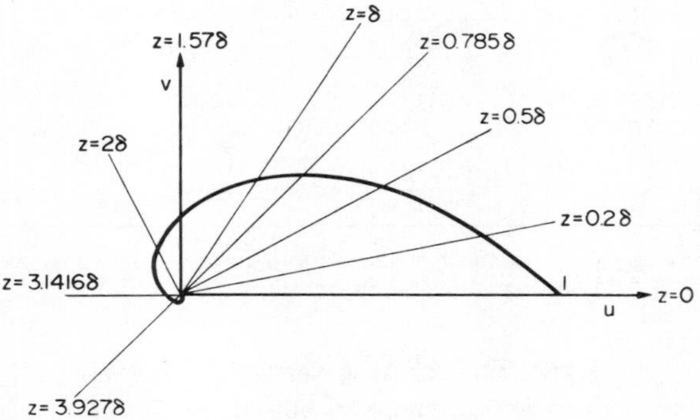

FIGURE 3.12 Ekman spiral (projection on the $z = 0$ plane) stretched coordinate, $\xi = z/\delta$.

Thus

$$\left\{\begin{aligned} \tilde{u} &= \frac{\tau\delta}{\sqrt{2}\,\rho\nu}\,\mathrm{e}^{-z/\delta}\cos(z/\delta-45^\circ) \\ \tilde{v} &= -\frac{\tau\delta}{\sqrt{2}\,\rho\nu}\,\mathrm{e}^{-z/\delta}\sin(z/\delta-45^\circ) \end{aligned}\right\} \tag{3.247}$$

It means that in the northern hemisphere the current is 45° to the right of the stress direction. One could have written

$$\begin{aligned} \tau_{xz} &= \rho\nu\left(\frac{\mathrm{d}\tilde{u}}{\mathrm{d}z}\right)_{z=0} \\ \tau_{yz} &= \rho\nu\left(\frac{\partial\tilde{v}}{\partial z}\right)_{z=0} \end{aligned} \tag{3.248}$$

and

$$\bar{\bar{\tau}} = -\rho\nu\left(\vec{i}\,\frac{\mathrm{d}\tilde{u}(0)}{\mathrm{d}z} + \vec{j}\,\frac{\mathrm{d}\tilde{v}(0)}{\mathrm{d}z}\right) \tag{3.249}$$

TABLE 3.1 *Calculation of an Ekman spiral*

$\zeta = E^{-1/2}z$		$e^{-\zeta}$	$u = e^{-\zeta}\cos\zeta$	$v = e^{-\zeta}\sin\zeta$
rad	deg			
0	0	1	1	0
0.10	5°729	0.9048	0.9003	0.0903
0.20	11°459	0.8187	0.8024	0.1627
0.40	22°918	0.6703	0.6174	0.2610
0.60	34°378	0.5488	0.4530	0.3099
0.7854	45°	0.4559	0.3224	0.3224
1.0	57°296	0.3679	0.1988	0.3096
1.40	80°021	0.2466	0.0419	0.2430
1.5708	90°	0.2079	0	0.2079
2.0	114°592	0.1353	−0.0563	0.1231
3.1416	180°	0.0432	−0.0432	0
3.927	225°	0.0197	−0.0139	−0.0139
4.712	270°	0.0090	0	−0.0090
5.498	315°	0.0041	−0.0029	+0.0029

The Ekman spiral is clearly an effect of the Coriolis force on the stress acting between the successive layers of fluid flowing with different velocities.

3.16 The rotating vertical cylinder: Stewartson sidewall boundary layers

We consider here a vertical cylinder rotating with a constant angular velocity Ω about its axis, filled with an incompressible viscous fluid. The flow being in a state of rigid body rotation, a small perturbation is produced by giving to the top (or bottom) surface a small additional velocity.

Starting with the Greenspan differential equation (3.93) in the case of a steady motion:

$$\left[E^2\nabla^6 + 4\frac{\partial^2}{\partial z^2}\right]p(r, \theta, z) = 0 \tag{3.250}$$

we have for the *horizontal boundaries*, in the boundary layer ($\partial/\partial z$ gradient dominating):

$$\left[E^2\frac{\partial^6}{\partial z^6} + 4\frac{\partial^2}{\partial z^2}\right]p(r, \theta, z) = 0 \tag{3.251}$$

which can be integrated two times and gives

$$\left[E^2\frac{\partial^4}{\partial z^4} + 4\right]p = zA(r, \theta) + B(r, \theta) \tag{3.252}$$

but as $p \to 0$ when z is large: $A = 0$, $B = 0$.

If we introduce a stretched coordinate defined by

$$\xi = E^a z \qquad \partial/\partial\xi = E^{-a}\,\partial/\partial z \tag{3.253}$$

we get

$$\left[E^2\,E^{4a}\frac{\partial^4}{\partial \xi^4} + 4\right]p(r, \theta, \xi) = 0 \tag{3.254}$$

and to ensure a proper balance between Coriolis and viscous forces one must have

$$2 + 4a = 0 \quad \text{or} \quad a = -\tfrac{1}{2}$$

this is an *Ekman layer of thickness of order* $E^{1/2}$.

Along the sidewall *vertical boundaries*, on the contrary, $\partial/\partial r$ is the dominating gradient and we have to write as boundary layer equation:

$$\left[E^2\frac{\partial^6}{\partial r^6} + 4\frac{\partial^2}{\partial z^2}\right]p = 0 \tag{3.255}$$

and introduce a stretched coordinate defined by

$$\xi = E^a(R - r) \tag{3.256}$$

with

$$\frac{\partial}{\partial \xi} = -E^{-a}\frac{\partial}{\partial r} \tag{3.257}$$

the equation becomes

$$\left[E^2\,E^{6a}\frac{\partial^6}{\partial \xi^6} + 4\frac{\partial^2}{\partial z^2}\right]p = 0 \tag{3.258}$$

the balance between Coriolis and viscous forces requires

$$2+6a=0 \quad \text{or} \quad a=-\tfrac{1}{3}$$

and we have a *Stewartson boundary layer of thickness of order* $E^{1/3}$, whose function is to reduce the vertical velocity to zero on the sidewall.

However it is shown that this layer, where the flow is strongly z-dependent, cannot match the flow in the interior which is z-independent and controlled by the Taylor–Proudman theorem (geostrophic) so that one needs, to satisfy all boundary conditions a *thicker layer of order* $E^{1/4}$. *This double structure on lateral boundaries is called "Stewartson layers"* (Barcilon, 1978).

The thick boundary layer ($E^{1/4}$) cannot satisfy all the boundary conditions on the wall while the thinner boundary layer ($E^{1/3}$) is not capable of matching correctly the interior geostrophic flow.

Note: with $\Omega = 7 \times 10^{-5}\,\mathrm{s}^{-1}$, $L = 2 \times 10^{8}$ cm, $\nu = 10^{-2}\,\mathrm{cm^2\,s^{-1}}$ one has $E = 3 \times 10^{-15}$, that is $E^{1/2} = 5.5 \times 10^{-7}$, $E^{1/3} = 1.45 \times 10^{-5}$, $E^{1/4} = 7.4 \times 10^{-3}$; thus $E^{1/2}L = 1$ m, $E^{1/3}L = 30$ m, $E^{1/4}L = 15$ km; i.e. a factor of 10,000 between $E^{1/4}$ and $E^{1/2}$.

Barcilon and Pedlosky (1967) show that in a rotating *stably stratified* fluid, vertical motions are inhibited, the Ekman layer suction no longer controls the interior dynamics, the Ekman layers themselves are frequently absent and the vertical Stewartson boundary layers are replaced by a new kind of boundary

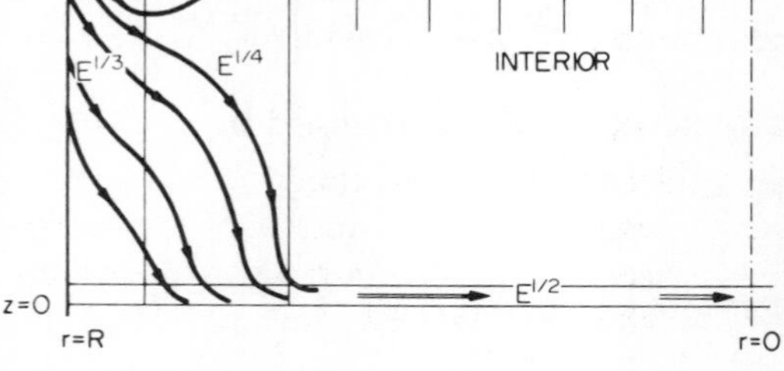

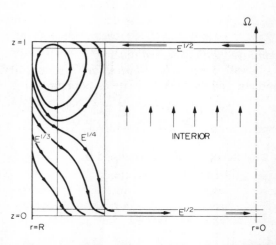

FIGURE 3.13 Sidewall boundary layers according to Barcilon (1971): An outer vertical boundary layer of thickness $E^{1/4}$ matches the interior flow while an inner vertical boundary layer of thickness $E^{1/3}$ ensures the zero velocity upon the vertical wall. These two embedded boundary layers are terminated at the top and at the bottom by a thinner Ekman layer of thickness $E^{1/2}$.

layer of thickness of $O(E^{1/2})$, whose structure is characteristic of rotating stratified fluids.

The interior dynamics are found to be controlled by dissipative processes.

3.17 Spin-up, Ekman suction

When one considers an infinitesimal change in the rotation speed of the container: $\Omega = \Omega_0(1+\varepsilon)$ the problem can be considered as linear and ε is the Rossby number. Thus:

$$\varepsilon = \frac{\Delta\Omega}{\Omega} \tag{3.259}$$

and is, in general, very small. But *spin-up from rest* is a strongly non-linear process, since in that case

$$\Delta\Omega = \Omega$$

and

$$\varepsilon = 1 \tag{3.260}$$

The limits of ε are

$0 < \varepsilon \leq 1$ in the spin-up case.
$-1 \leq \varepsilon < 0$ in the spin-down case.

When spin-up of the container occurs, the core of the fluid body remains initially unaffected but an Ekman boundary layer is immediately developed on the interface where the fluid begins to rotate more rapidly because of the viscous stress. A radial outflow results, which is due to the increased centrifugal action not balanced by the pressure gradients which remain practically unchanged. To compensate this flow, for reasons of continuity, an axial flow (orthogonal to the boundary) will bring fluid from the interior inside the boundary layer: this is the *Ekman suction* which implies a radial inflow in the core of the fluid. As the angular momentum of fluid rings moving inwards to replace the fluid entering in the Ekman layer must be conserved, the internal vorticity must increase: this is the *inertial* (and not diffusive) mechanism whereby the viscous boundary layer transmits information to the bulk of the fluid.

In rotating cylinders this circulation is closed by the Stewartson sidewall boundary layers parallel to $\overline{\Omega}$. The suction in the Ekman layer limits the amplitude in case of resonance. Weak internal modes (frequency 2Ω) are excited during this process. The "*spin-up time*" is the time required for the interior flow to approach a new final state of rigid rotation within e^{-1}. It is defined as

$$T_S = \frac{L}{\sqrt{\nu\Omega}} \tag{3.261}$$

Thus, for a weakly viscous liquid, a very long time is required for the bulk of the liquid to reach full spin. But the boundary layer is responsible for a secondary flow which convects spinning fluid from the boundary into the interior and, as a consequence, the fluid attains rotational motion many times faster than without secondary flow.

One should point out here that a *stable stratification inhibits vertical motions* and consequently Ekman suction. The Ekman layer, in that case, is a passive layer, while the Stewartson vertical boundary layers are replaced by a new kind of boundary layers of thickness $O(E^{1/2})$ (Barcilon and Pedlosky, 1967).

Note: a change is the *direction* of the axis of rotation gives rise to what one calls "*spin-over*".

Experiences conducted in Spacelab under microgravity conditions could help to solve many of the difficulties we meet in ground based hydrodynamics experiments where the imposed uniform gravity acceleration distorts the flow patterns in ways that do not correspond to actual planetary flows.

Radial gravity and the resulting buoyancy can be simulated in space with spherical fluid models by using electric radial forces which are proportional to the density of the dielectric fluid used (GFFC: Geophysical Fluid Flow Cell (*) Hart 1984).

3.18 Thermal Instability – The Rayleigh number.

Thermal instability in a non rotating horizontal incompressible fluid sheet, heated from below $\left(\frac{\partial T}{\partial x} = \frac{\partial T}{\partial y} = 0, \beta = -\frac{\partial T}{\partial z} \text{ constant}\right)$ can be analysed along the same lines as §3.7 or 3.12. By using the definition (2.26) and the Boussinesq approximation (3.68) where $\rho' = -\rho_0\, \alpha T$, the Navier equations, without the advection term, can be written:

$$\left.\begin{aligned} \frac{\partial u}{\partial t} &= -\frac{\partial p}{\partial x} + \nu\nabla^2 u \\ \frac{\partial v}{\partial t} &= -\frac{\partial p}{\partial y} + \nu\nabla^2 v \\ \frac{\partial w}{\partial t} &= g\alpha T - \frac{\partial p}{\partial z} + \nu\nabla^2 w \end{aligned}\right\} \tag{3.262}$$

with

$$\frac{\partial u}{\partial x} + \frac{\partial v}{\partial y} + \frac{\partial w}{\partial z} = 0$$

(*) The conditions of this NASA experience are: a spherical shell with internal and external radii: 24 and 33 mm, thus thickness of the shell 9 mm, periods of rotation between 1 and 255 s, Ekman number from 0.01 to 0.001, silicone oil with Prandtl number $P = 7.0$.

and the heat conduction equation (3.48):

$$\frac{dT}{dt} = \frac{\partial T}{\partial t} + w\frac{\partial T}{\partial z} = \chi\nabla^2 T$$

Or

$$\left(\frac{\partial}{\partial t} - \chi\nabla^2\right)T = \beta w \tag{3.263}$$

Taking, as previously, the curl in eqn. (3.262) one eliminates the gradient of p to obtain:

$$\left.\begin{aligned} \frac{\partial}{\partial t}\left(\frac{\partial w}{\partial y} - \frac{\partial v}{\partial z}\right) &= g\alpha\frac{\partial T}{\partial y} + \nu\nabla^2\left(\frac{\partial w}{\partial y} - \frac{\partial v}{\partial z}\right) \\ \frac{\partial}{\partial t}\left(\frac{\partial u}{\partial z} - \frac{\partial w}{\partial x}\right) &= -g\alpha\frac{\partial T}{\partial x} + \nu\nabla^2\left(\frac{\partial u}{\partial z} - \frac{\partial w}{\partial x}\right) \end{aligned}\right\} \tag{3.264}$$

We derive the first with respect to y and the second with respect to x and subtract, using the incompressibility condition:

$$\begin{aligned} \left(\frac{\partial}{\partial t} - \nu\nabla^2\right)\nabla^2 w &= g\alpha\nabla_H^2 T \\ \nabla_H^2 &= \frac{\partial^2}{\partial x^2} + \frac{\partial^2}{\partial y^2} \end{aligned} \tag{3.265}$$

Applying the $\left(\frac{\partial}{\partial t} - \chi\nabla^2\right)$ operator to eqn. 3.265 and taking eqn. 3.263 into account, one has

$$\left[\left(\frac{\partial}{\partial t} - \nu\nabla^2\right)\left(\frac{\partial}{\partial t} - \chi\nabla^2\right)\nabla^2 - \alpha\beta g\nabla_H^2\right]w = 0 \tag{3.266}$$

If w is independent from time t (stable convection), this equation reduces to:

$$\left(\nabla^6 - \frac{\alpha\beta g}{\chi\nu}\nabla_H^2\right)w = 0 \tag{3.267}$$

which obviously has, as characteristic dimensionless number:

$$Ra = \frac{\alpha\beta g}{\chi\nu}L^4 \tag{3.268}$$

the Rayleigh number.

When this dimensionless number becomes higher than a critical value (1708 in the simple case considered here) the convection becomes unstable. The thermodynamical significance of this phenomenon is described as follows by Chandrasekhar (1961, page 34):

"*Instability occurs at the minimum temperature gradient at which a balance can be steadily maintained between the kinetic energy dissipated by viscosity and the internal energy released by buoyancy force*".

The whole book of Chandrasekhar is devoted to hydrodynamic and hydromagnetic stability and instability.

The Rayleigh number is quite high in the Lower Mantle of the Earth: 3.2×10^5 if $\nu = 10^{24}$ stokes and 3.2×10^8 if $\nu = 10^{21}$ stokes, this latter value being likely in the D'' thermal boundary layer at the bottom of the mantle which therefore looks unstable.

The ratio of the Rayleigh number to the Prandtl number is called Grashof number

$$Gr = \frac{g\beta U L^3}{V^2} \tag{3.269}$$

and expresses the ratio between buoyancy and sluggishness due to viscosity.

3.19 An application of Navier equations and boundary layer theory to the problem of differential rotation speed of the core and mantle of the Earth

Bondi and Lyttleton (1948) and I. Proudman (1956) have used the Navier equations to determine the flow in the liquid core of the Earth when the mantle and the core have different rotation speed.

The basic equations are, of course, the same in both papers but Bondi and Lyttleton use them to investigate the effect of the secular retardation of the mantle (proportional to the time) on the flow in a spherical liquid core without solid inner core, while Proudman considers a constant difference of speed but introduces an inner boundary at the inner-outer core interface.

C. E. Pearson (1967) proceeded by numerical integrations to describe the flow in a spherical shell but has not considered the essential Proudman result about a Stewartson sidewall cylindrical boundary layer, tangent to the inner core.

Let us consider a fluid envelope bounded by two rigid spherical concentric surfaces of radii $r = a$ (inner core), $r = \alpha a$ (mantle).

The flow resulting from a small differential rotation may be considered as a small perturbation superimposed upon a rigid body rotation. This allows to linearize the governing equations in the magnitude of the perturbation.

The equation of continuity, written in spherical coordinates is the equation (3.25). Because of the rotational symmetry of the system all dynamical variables must be independent of the longitude and one may apply eqn. (3.26) and integrate the equation in terms of a stream function as given by the equation (3.27) in spherical coordinates:

$$u = \frac{1}{r^2 \sin\theta}\frac{\partial\psi}{\partial\theta}, \quad v = -\frac{1}{r\sin\theta}\frac{\partial\psi}{\partial r} \tag{3.270}$$

since $\dfrac{\partial w}{\partial\lambda} = 0$ (eqn. 3.26).

The solution for the azimuthal velocity is of the form

$$w = f(r\sin\theta)$$

or, as written by Proudman

$$w = \frac{\chi}{r\sin\theta} \tag{3.271}$$

The boundary conditions are:
on the inner sphere $r = a$:

$$\frac{\partial\psi}{\partial r} = \psi = 0 \qquad \text{as } u = v = 0$$

$$w = \Omega a \sin\theta$$

thus

$$\chi = \Omega\,(a^2 \sin^2\theta)$$

on the outer sphere $r = \alpha a$

$$\frac{\partial\psi}{\partial r} = \psi = 0 \qquad \text{as } u = v = 0$$

$$w = \Omega(1+\varepsilon)\alpha a \sin\theta$$

thus

$$\chi = \Omega\,(1+\varepsilon)\alpha^2 a^2 \sin^2\theta$$

When ε is zero, one has a rigid rotation. Therefore, as ε is *small* one can propose to write, inside the fluid body:

$$\left.\begin{array}{l} \text{for } \psi\colon\ \varepsilon\Omega a^3\,\psi \\ \text{for } \chi\colon\ \Omega r^2 \sin^2\theta + \varepsilon\Omega a^2 \chi \end{array}\right\} \tag{3.272}$$

because u, v are proportional to $\varepsilon\Omega a$.

Let us take, from now, r for r/a so that

$$\left.\begin{array}{l} u = \varepsilon\Omega a \dfrac{\partial\psi}{r^2 \sin\theta\partial\theta} \\[2ex] v = -\varepsilon\Omega a \dfrac{\partial\psi}{r\sin\theta\partial r} \\[2ex] w = \Omega a r \sin\theta + \varepsilon\Omega \dfrac{a\chi}{r\sin\theta} \end{array}\right\} \tag{3.273}$$

The Navier equations are:

$$\left.\begin{aligned}\left(\frac{\partial}{\partial t}-\nu D^2\right)D^2\psi+2\Omega\left(\frac{\partial\chi}{\partial r}\cos\theta-\frac{\partial\chi}{r\partial\theta}\sin\theta\right)=0\\ \left(\frac{\partial}{\partial t}-\nu D^2\right)\chi-2\Omega\left(\frac{\partial\psi}{\partial r}\cos\theta-\frac{\partial\psi}{r\partial\theta}\sin\theta\right)=0\end{aligned}\right\}\quad(3.274)$$

with

$$D^2=\frac{\partial^2}{\partial r^2}+\frac{\sin\theta}{r^2}\frac{\partial}{\partial\theta}\left(\frac{1}{\sin\theta}\frac{\partial}{\partial\theta}\right)$$

and the boundary conditions:

$$\left.\begin{aligned}&\text{inner sphere}\quad r=1\quad \frac{\partial\psi}{\partial r}=\psi=0,\quad \chi=a^2\sin^2\theta\Omega\\ &\text{outer sphere}\quad r=\alpha\quad \frac{\partial\psi}{\partial r}=\psi=0,\quad \chi=(\alpha^2a^2\sin^2\theta)\Omega(1+\varepsilon)\end{aligned}\right\}\quad(3.275)$$

Choosing cylindrical coordinates substantially simplifies these equations making them more appropriate than the spherical coordinates. With

$$\rho=r\sin\theta\qquad \zeta=r\cos\theta\quad(3.276)$$

and

$$\left.\begin{aligned}\frac{\partial}{\partial r}&=\frac{\partial}{\partial\zeta}\frac{\partial\zeta}{\partial r}+\frac{\partial}{\partial\rho}\frac{\partial\rho}{\partial r}=\cos\theta\frac{\partial}{\partial\zeta}+\sin\theta\frac{\partial}{\partial\rho}\\ \frac{\partial}{r\partial\theta}&=\frac{\partial}{\partial\zeta}\frac{\partial\zeta}{r\partial\theta}+\frac{\partial}{\partial\rho}\frac{\partial\rho}{r\partial\theta}=-\sin\theta\frac{\partial}{\partial\zeta}+\cos\theta\frac{\partial}{\partial\rho}\end{aligned}\right\}\quad(3.277)$$

The Navier equations become indeed:

$$\begin{aligned}\left(\frac{\partial}{\partial t}-\nu D^2\right)D^2\psi+2\Omega\frac{\partial\chi}{\partial\zeta}=0\\ \left(\frac{\partial}{\partial t}-\nu D^2\right)\chi-2\Omega\frac{\partial\psi}{\partial\zeta}=0\end{aligned}\quad(3.278)$$

with

$$D^2=\frac{\partial^2}{\partial\rho^2}-\frac{\partial}{\rho\partial\rho}+\frac{\partial^2}{\partial\zeta^2}$$

{equations (20–21) of Bondi Lyttleton; equations (2.14–16) of Proudman}.

Moreover, when using the cylindrical coordinates, we may take advantage of the symmetry of the two hemispheres which imposes that χ will be an even function of ζ while ψ will be an odd function of ζ. This makes

$$u^* = \partial\psi/\rho\partial\rho, \qquad \text{velocity in } \zeta \text{ direction, an odd function}$$

$$v^* = -\partial\psi/\rho\partial\zeta, \qquad \text{velocity in } \rho \text{ direction, an even function}$$

Then, if one wants to introduce the secular retardation of the Earth's rotation and obtain the Bondi-Lyttleton equations, one has simply to replace the function χ by

$$k\rho^2 t + \chi \qquad \text{with } k < 0 \tag{3.279}$$

in the second equation (3.278) which gives

$$\nu D^2\chi + 2\Omega \frac{\partial\psi}{\partial\zeta} = k\rho^2 \tag{3.280}$$

In the bulk of the fluid the viscous term can be neglected (it is important only in the boundary layers) when ν is as small as in the Earth's core, so that

$$\psi = \frac{k}{2\Omega}\rho^2\zeta \tag{3.281}$$

and, from the first equation (3.278):

$$\frac{\partial\chi}{\partial\zeta} = 0 \quad \text{gives} \quad \chi = \chi(\rho) \tag{3.282}$$

It results that

$$u^* = \frac{k}{\Omega}\zeta \qquad v^* = -\frac{k}{2\Omega}\rho \tag{3.283}$$

Thus, if k is negative (retardation of the mantle), the flow is away from the axis of rotation and towards the equator

The boundary layer equations

We will now consider the development of the boundary layer equations according to Proudman, considering that the two boundaries have a constant but different speed of rotation.

The Navier equations become:
in spherical coordinates

$$\left.\begin{aligned} 2\left(\frac{\partial\chi}{\partial r}\cos\theta - \frac{\partial\chi}{r\partial\theta}\sin\theta\right) &= ED^4\psi \\ -2\left(\frac{\partial\psi}{\partial r}\cos\theta - \frac{\partial\psi}{r\partial\theta}\sin\theta\right) &= ED^2\chi \end{aligned}\right\} \tag{3.284}$$

in cylindrical coordinates

$$\left.\begin{aligned} 2\frac{\partial \chi}{\partial \zeta} &= ED^4\psi \\ -2\frac{\partial \psi}{\partial \zeta} &= ED^2\chi \end{aligned}\right\} \tag{3.285}$$

(with the respective expressions of D^2 given above)

E is the Ekman number.

When E is very small the equations (3.285) give

$$\frac{\partial \chi}{\partial \zeta} \sim 0 \qquad \frac{\partial \psi}{\partial \zeta} \sim 0$$

so that, apart from singular surfaces, one can put

$$\chi = \chi(\rho) \qquad \psi = \psi(\rho)$$

which means that the velocity depends only from the distance to the axis of rotation and not from the distance to the equator (like Taylor columns). Such solutions cannot satisfy the boundary conditions on the two spherical boundaries.

One can suppose that singular surfaces take the form of *conventional* boundary layers which means that $\dfrac{\partial}{r\partial\theta}$ can be neglected with respect to $\dfrac{\partial}{\partial r}$. This is the classical boundary layer approximation.

With this approximation the equations (3.284) inside the boundary layer become ordinary differential equations:

$$\begin{aligned} 2\frac{\partial \chi}{\partial r}\cos\theta &= E\frac{\partial^4\psi}{\partial r^4} \\ -2\frac{\partial \psi}{\partial r}\cos\theta &= E\frac{\partial^2\chi}{\partial r^2} \end{aligned} \tag{3.286}$$

One integration gives

$$\left.\begin{aligned} 2\cos\theta\,(\chi - \chi_0) &= E\frac{\partial^3\psi}{\partial r^3} \\ -2\cos\theta\,(\psi - \psi_0) &= E\frac{\partial \chi}{\partial r} \end{aligned}\right\} \tag{3.287}$$

where χ_0 and ψ_0 are functions of θ *only*.

Combining with (3.286) gives

$$\frac{\partial^4}{\partial r^4}(\psi-\psi_0)+\frac{4\cos^2\theta}{E^2}(\psi-\psi_0)=0 \tag{3.288}$$

which is the classical Ekman equation (3.236).

Inner boundary $\rho = \sin\theta$

Let us use now a stretched coordinate defined as

$$\eta=(r-1)(E^{-1}\cos\theta)^{1/2} \tag{3.289}$$

The four independent solutions of eqn. (3.288) are

$$\exp(\pm 1 \pm i)\eta$$

or

$$\psi=\psi_0\{1-e^{-\eta}(\cos\eta+\sin\eta)\} \tag{3.290}$$

and, consequently from the first eqn. (3.287) where

$$\frac{\partial^3\psi}{\partial r^3}\rightarrow\eta^3 \;\therefore\; (E^{-1}\cos\theta)^{3/2}$$

one obtains

$$\chi=\chi_0-2(E^{-1}\cos\theta)^{1/2}\psi_0\, e^{-\eta}\cos\eta$$

As χ must vanish at $r=1$, $\eta=0$, one has also

$$\chi_0=2(E^{-1}\cos\theta)^{1/2}\psi_0 \tag{3.291}$$

and finally

$$\chi=2(E^{-1}\cos\theta)^{1/2}\psi_0(1-e^{-\eta}\cos\eta) \tag{3.292}$$

Note that, according to eqn. (3.289) the boundary layer is thickening towards the equator where it breaks down.

It also results that eqn. (3.291) implies a relation between the functions ψ and χ throughout the whole *of that part* of the core of the motion *which lies within the cylinder* $\rho = 1$, *tangent to the inner sphere.* Thus, from (3.291):

$$\chi_0(\rho)=2E^{-1/2}(1-\rho^2)^{1/4}\psi_0(\rho) \quad \text{for} \quad \rho<1 \tag{3.292}$$

Outer boundary ($\rho = \alpha\sin\theta$)

The integration of eqn. (3.286) gives here

$$\left.\begin{aligned} 2\cos\theta\{\chi-\chi_0(\alpha\sin\theta)\} &= E\frac{\partial^3\psi}{\partial r^3} \\ -2\cos\theta\{\psi-\psi_0(\alpha\sin\theta)\} &= E\frac{\partial\chi}{\partial r} \end{aligned}\right\} \tag{3.293}$$

and the stretched variable to be taken is

$$\eta' = (\alpha - r)(E^{-1}\cos\theta)^{1/2}$$

with the solutions

$$\left.\begin{aligned} \psi &= \psi_0(\alpha\sin\theta)\{1 - e^{-\eta'}(\cos\eta' + \sin\eta')\} \\ \chi &= \chi_0(\alpha\sin\theta) + 2(E^{-1}\cos\theta)^{1/2}\psi_0(\alpha\sin\theta)e^{-\eta'}\cos\eta' \end{aligned}\right\} \tag{3.294}$$

The boundary condition requires that

$$\chi = \alpha^2\sin^2\theta \quad \text{over } r = \alpha, \text{ that is } \eta' = 0$$

Thus

$$\alpha^2\sin^2\theta - \chi_0(\alpha\sin\theta) = 2(E^{-1}\cos\theta)^{1/2}\psi_0(\alpha\sin\theta) \tag{3.295}$$

a relation which must be satisfied throughout the whole core. This may be written

$$\rho^2 - \chi_0(\rho) = 2E^{-1/2}\left(1 - \frac{\rho^2}{\alpha^2}\right)^{1/4}\psi_0(\rho) \tag{3.296}$$

without restriction on ρ $\left(\cos^2\theta = -\frac{\rho^2}{\alpha^2}\right)$.

Equations (3.292) and (3.296) uniquely determine the functions ψ_0 and χ_0 in the region of the cylinder $\rho < 1$.

$$\left.\begin{aligned} \psi_0(\rho) &= \frac{\rho^2}{2}E^{1/2}\left\{\left(1 - \frac{\rho^2}{\alpha^2}\right)^{1/4} + (1-\rho^2)^{1/4}\right\}^{-1} \\ \chi_0(\rho) &= \rho^2(1-\rho^2)^{1/4}\left\{\left(1 - \frac{\rho^2}{\alpha^2}\right)^{1/4} + (1-\rho^2)^{1/4}\right\}^{-1} \end{aligned}\right\} \tag{3.297}$$

It is interesting to point out that χ_0 results from the viscosity but do not depend upon E.

The conclusions of Proudman are that "In the core of the motion the streamlines in an axial plane are parallel to the axis of rotation, so that fluid leaving the boundary layer on one sphere round the circle $\rho = \rho_0$ must enter the boundary layer on the other sphere round a circle of the same radius. The other essential connection arises from the fact that the azimuthal velocity in the core is also a function only of the radial distance from the axis of rotation."

It follows from (3.297) that the cylindrical surface $\rho = 1$ must be another singular surface of the motion.

Applying to the equations (3.285) the Stewartson procedure (§3.16), Proudman obtains a cylindrical boundary layer, parallel to the axis of rotation and tangent to the inner core, with a thickness proportional to $E^{1/3}$.

Bibliography

Aldridge, K. D. (1972) Axisymmetric inertial oscillations of a fluid in a rotating spherical shell. *Mathematika* **19**, 163–8.

Aldridge, K. D. (1975) Inertial waves and the Earth's outer core. *Geophys. J. R. Astr. Soc.*, **42**, 337–45.

Aldridge, K. D. and Toomre, A. (1969) Axisymmetric inertial oscillations of a fluid in a rotating spherical container. *J. Fluid Mech.*, **37**(2), 307–23.

Allan, R. R. (1978) *Boundary layers in Homogeneous-Rotating Fluids.* Part I—Notes on lectures. The University Press of Florida, Gainesville, 1–68.

André, J. C. and Barrere, M. (1982) *Turbulence Fluide. Ecoulements Complexes.* Ecole Polytechniquè, Paris, 2 Fascicules.

Backus, G. (1968) Kinematics of geomagnetic secular variation in a perfectly conducting core. *Phil. Trans. Roy. Soc. Lond.*, **A263**, 239–66.

Baines, P. G. (1967) Forced oscillations of an enclosed rotating fluid *J. Fluid Mech.*, **30**(3), 533–46.

Barcilon, V. (1978) *Boundary Layers in Stratified-rotating Fluids* (ed. J. S. Fein). The University Press of Florida, 69–124.

Barcilon, V. and Pedlosky, J. (1967) Linear theory of rotating stratified fluid motions. *J. Fluid Mech.*, **29**(1), 1–16.

Bartels, F. (1982) Taylor vortices between two concentric rotating spheres *J. Fluid Mech.*, **119**, 1–25.

Bateman, H. (1929) Notes on a differential equation which occurs in the two-dimensional motion of a compressible fluid and the associated variational problems. *Proc. Roy Soc. Lond.*, **A125**, 598–618.

Benton, E. R. and Clark, A. Jr. (1974) Spin-up. *Ann. Rev. Fluid Mech.*, **6**, 257–79.

Bondi, H. and Lyttleton, R. A. (1948, 1953) On the dynamical theory of the rotation of the Earth. I: The secular retardation of the core. *Proc. Camb Phil. Soc.*, **44**, 345–59; II: The effect of precession on the motion of the liquid core. *Proc. Camb. Phil. Soc.*, **49**, 498–515.

Booker, J. R. and Bretherton, F. P. (1967) The critical layer for internal gravity waves in a shear flow. *J. Fluid Mech.*, **27**, 513–39.

Bretherton, F. P., Carrier, G. F. and Longuet-Higgins, M. S. (1966) Report on the IUTAM symposium on rotating fluid systems. *J. Fluid Mech.*, **26**, 393.

Bryan, G. H. (1888) The waves on a rotating liquid spheroid of finite ellipticity. *Phil. Trans. Roy. Soc. Ser. A*, **180**, 187–219.

Busse, F. H. (1968a) Steady fluid flow in a precessing spheroidal shell. *J. Fluid Mech.*, **33**(4), 738–51.

Busse, F. H. (1968b) Shear flow instabilities in rotating systems. *J. Fluid Mech.*, **33**(3), 577–89.

Busse, F. H. (1970a) Thermal instabilities in rapidly rotating systems. *J. Fluid Mech.*, **44**(3), 441–60.

Busse, F. H. (1970b) Differential rotation in stellar convection zones. *Astrophys. J.*, **159**, 629–39.

Busse, F. H. (1971) Bewegungen im kern der erde. *Z. Geophys.*, Band 37. Physica-Verlag, Wurzburg, 153–77.

Busse, F. H. (1973) Differential rotation in stellar convection zones. II. *Astron. Astrophys.*, **28**, 27–37.

Busse, F. H. (1975) Patterns of conversion in spherical shells. *J. Fluid Mech.*, **72**, 65–85.

Busse, F. H. and Carrigan, C. R. (1974) Convection induced by centrifugal buoyancy. *J. Fluid Mech.*, **62**, 579–92.

Busse, F. H. and Carrigan, C. R. (1976) Laboratory simulation of thermal convection in rotating planets and stars. *Science*, **191**, 81–3.

Carrier, G. F. (1965) Some effects of stratification and geometry in rotating fluids. *J. Fluid Mech.*, **23**, 145–72.

Carrigan, C. R. and Gubbins, D. (1979) The source of the Earth's magnetic field. *Sci. Am.*, **240**, 92–101.

Cartan, E. (1922) Sur les petites oscillations d'une masse fluide. *Bull. Sci. Math.*, **XLVI**, 317–69.

Crossley, D. (1984) Oscillatory flow in the liquid core. *Phys. Earth Plan.* Int., **36**, 1–16.

Crossley, D. J. (1975a) Core undertones with rotation. *Geophys. J. R. Astr. Soc.*, **42**, 477–88.

Crossley, D. J. (1975b) The free-oscillation equations at the centre of the earth. *Geophys. J. R. Astr. Soc.*, **41**, 153–63.

Crossley, D. J. and Rochester, M. G. (1980) Simple core undertones. *Geophys. J. R. Astr. Soc.*, **60**, 129–61.

Ekman, V. M. (1905) On the influence of the Earth's rotation on ocean-currents. *Arkiv Matem., Astron. Och Fysik*, **2**(11), 53.
Fadnis, B. S. (1954) Boundary layer on rotating spheroids. *Z. Angew. Math. Phys.* **5**, 156–63.
Fearn, D. R. and Loper, D. E. (1981) Compositional convection and stratification in the Earth's core. *Nature*, **289**, 393–394.
Fowler, A. C. (1985) A simple model of convection in the terrestrial planets. *Geophys. Astrophys. Fluid Dyn.*, **31**, 283–309.
Friedlander, S. (1982) Turning surface behaviour for internal waves subject to general gravitational fields. *Geophys. Astr. Fluid. Dyn.*, **21**, 189–200.
Friedlander, S. (1985) Internal oscillations in the Earth's Fluid core. *Geophys. J. R. Astr. Soc.*, **80**, 345–61.
Friedlander, S. (1985) Stability of the subseismic wave equation for the Earth's fluid core. *Geophys. Astrophys. Fluid Dyn.*, **31**, 151–67.
Friedlander, S. and Siegmann, W. L. (1982a) Internal waves in a rotating stratified fluid in an arbitrary gravitational field. *Geophys. Astr. Fluid Dyn.*, **19**, 267–91.
Friedlander, S. and Siegmann, W. L. (1982b) Internal waves in a contained rotating stratified fluid. *J. Fluid Mech.*, **114**, 123–56.
Friedlander, S. and Siegmann, W. L. (1983) Effects of dissipation on internal waves in a contained rotating stratified fluid. *Geophys. Astr. Fluid. Dyn.*, **27**, 183–216.
Gans, R. F. (1970) On the precession of a resonant cylinder. *J. Fluid Mech.*, **41**(4), 865–72.
Greenhill, A. G. (1880) On the general motion of a liquid ellipsoid under the gravitation of its own parts. *Procs. Camb. Phil. Soc.*, **4**, 4–14.
Greenspan, H. P. (1964) On the transient motion of a contained rotating fluid. *J. Fluid Mech.*, **20**(4), 673–96.
Greenspan, H. P. (1965) On the general theory of contained rotating fluid motions. *J. Fluid Mech.*, **22**(3), 449–62.
Greenspan, H. P. and Howard, L. N. (1963) On a time-dependent motion of a rotating fluid. *J. Fluid Mech.*, **17**, 385–404.
Greenspan, H. and Weinbaum, S. (1965) On non-linear spin-up of a rotating fluid. *J. Math. Phys.* **44**, 66–85.
Gubbins, D., Thomson, C. J. and Whaler, K. A. (1982) Stable regions in the Earth's liquid core. *Geophys. J. R. Astr. Soc.*, **68**, 241–51.
Hadamard, J. (1936) Conférences internationales sur les équations aux dérivées partielles. Conditions propres à déterminer les solutions. *L'Enseignement Math.*, **35**, 5.
Hide, R. (1956) The hydrodynamics of the Earth's core. *Physics and Chemistry of the Earth*, Pergamon Press, Vol. 1, 94–137.
Hide, R. (1966) Free hydromagnetic oscillations of the Earth's core and the theory of the geomagnetic secular variation *Phil. Trans. Roy. Soc. Lond.* **A259**, 615–47.
Hide, R. (1978) Dynamics of rotating fluids. *Rotating fluids in geophysics*, Academic Press, 1–26.
Hide, R. (1981) The magnetic flux linkage of a moving medium: a theorem and geophysical applications. *J. Geophys. Res.*, **86**, 11,681–7.
Hide, R. and Ibbetson, A. (1966) An experimental study of Taylor columns. *Icarus*, **5**, 279–90.
Hide, R. and Stewartson, K. (1972) Hydromagnetic oscillations of the Earth's core. *Rev. Geophys. Space Phys.*, **10**(2), 579–98.
Hough, S. S. (1895) The oscillations of a rotating ellipsoidal shell containing fluid. *Phil. Trans. Roy. Soc. Lond.*, **186**(1), 469–506.
Howarth, L. (1951a) The boundary layer in three-dimensional flow—Part I: derivation of the equations for flow along a general curved surface. *Phil. Mag.*, **42**, 239–43.
Howarth, L. (1951b) Note on the boundary layer on a rotating sphere. *Phil. Mag.*, **42**, 1308–14.
Ibbetson, A. (1967) Some laboratory experiments on Rossby waves in a rotating annulus. *Tellus*, **19**, 81–6.
Illingworth, C. R. (1953a) The laminar boundary layer of a rotating body of revolution. *Phil. Mag.*, Ser. 7, **44**(351), 389–403.
Illingworth, C. R. (1953b) Boundary layer growth on a spinning body. *Phil. Mag.*, Ser. 7, **45**, 1–8.
Kelvin, W. (1885) On the motion of a liquid within an ellipsoidal hollow. *Kelvin Math, Phys. Papers.* **IV**, 193–201; (1910) *Proc. Roy. Soc. Edinburgh*, **XIII**, 370–8.
Kovasznay, S. G. (1970) The turbulent boundary layer. *Ann. Rev. Fluid Mech.*, **2**, 95–109.
Lighthill, M. J. (1966) Dynamics of rotating fluids: a survey. *J. Fluid Mech.*, **26**(2), 411–31.

London, S. and Shen, M. C. (1979) Free oscillation in a rotating spherical shell. *Phys. Fluids*, **22**(11), 2071–80.

Longuet-Higgins, M. S. (1964) Planetary waves on a rotating sphere. *Proc. Roy. Soc. Lond.*, **A279**, 446–73.

Longuet-Higgins, M. (1965) Planetary waves on a rotating sphere. II. *Proc. Roy. Soc. Lond.* **284**, 40–68.

Longuet-Higgins, M. S. (1966) Planetary waves on a hemisphere bounded by meridians of longitude. *Phil. Trans. Roy. Soc. Lond.* **260**, 318–50.

Longuet-Higgins, M. (1968) The Eigenfunctions of Laplace tidal equations over a sphere. *Phil. Trans. Roy. Soc. Lond.*, **A262**(1132), 511–607.

Malkus, W. V. R. (1971) Motions in the fluid core. *Mantello e Nucleo Nella Fisica Planetaria. Scuola Int. di Fisica, E. Fermi*, 38–63, Academic Press.

Malkus, W. V. R. (1973) Convection at the melting point: a thermal history of the Earth's core. *Geophys. Fluid Dyn.*, **4**, 267–78.

McEwan, A. D. (1970) Inertial oscillations in a rotating fluid cylinder. *J. Fluid Mech.*, **40**(3), 603–40.

Miles, J. W. (1974) On Laplace's tidal equations. *J. Fluid Mech.*, **66**(2), 241–60.

Moore, D. W. (1978) Homogeneous fluids in rotating-viscous effects. *Rotating Fluids in Geophysics.* Academic Press, 29–65.

Munk, W. H. (1980) Internal wave spectra at the buoyant and inertial frequencies. *J. Phys. Oceanogr.* **10**, 1718–28.

Musen, P. (1978) On the tidal oscillations of the liquid core of the Earth. *NASA Technical Paper* 1223, 1–61.

Nakabayashi, K., Yamada, Y. and Kishimoto, T. (1982) Viscous frictional torque in the flow between two concentric rotating rough cylinders. *J. Fluid Mech.*, **119**, 409–22.

Needler, G. T. and Leblond, P. H. (1973) On the influence of the horizontal component of the Earth's rotation on long period waves. *Geophys. Fluid Dyn.*, **5**, 23–46.

Nigam, S. D. (1954) Note on the boundary layer on a rotating sphere. *Z. Angew. Math. Phys.*, **5**, 151–5.

Olson, P. (1977) Internal waves in the Earth's core. *Geophys. J. R. Astr. Soc.*, **51**, 183–215.

Pearson, Carl E. (1967) A numerical study of the time-dependent viscous flow between two rotating spheres. *J. Fluid Mech.*, **28**(2), 323–36.

Pedlosky, J. (1979) *Geophysical Fluid Dynamics.* Springer Verlag.

Pekeris, C. L. and Accad, Y. (1972) Dynamics of the liquid core of the Earth. *Phil. Trans. Roy. Soc. Lond.*, **A273**, 237–60.

Poincaré, H. (1885) Sur l'équilibre d'une masse fluide animée d'un mouvement de rotation. *Acta Math.*, **7**, 259–80.

Poincaré, H. (1910) Sur la précession des corps déformables. *Bull. astr.* **XXVII**, 322–56.

Proudman, J. (1942a) On Laplace's differential equations for the tides. *Proc. Roy. Soc.* **A179**(978), 261–88.

Proudman, J. (1942b) On Laplace's differential equations for the tides. *Math. Rev.*, **3**, 286.

Proudman, I. (1956) The almost-rigid rotation of viscous fluid between concentric spheres. *J. Fluid Mech.*, **1**, 505–16.

Riahi, N., Geiger, G. and Busse, F. H. (1982) Finite Prandtl number convection in spherical shells. *Geophys. Astr. Fluid Dyn.*, **20**, 307–18.

Rickard, J. A. (1973) Free oscillations of a rotating fluid contained between two spheroidal surfaces. *Geophys. Fluid Dyn.*, **5**, 369–83.

Roberts, P. H. (1968) On the thermal instability of a rotating-fluid sphere containing heat sources. *Phil. Trans. Roy. Soc., Lond.*, **A263**(1136), 93–117.

Roberts, P. H. and Stewartson, K. (1963) On the stability of a Maclaurin spheroid of small viscosity. *Astr. J.*, **137**, 777–90.

Roberts, P. H. and Stewartson, K. (1965) On the motion of a liquid in a spheroidal cavity of a precessing rigid body. II. *Proc. Camb. Phil. Soc.*, **61**, 279–88.

Robinson, A. R. (1978) *Boundary Layers in Homogeneous-Rotating Fluids* (ed. J. S. Fein). The University Press of Florida, 1–68.

Rott, N. and Rosenzweig, M. L. (1960) On the response of the laminar boundary layer to small fluctuations of the free-stream velocity. *J. Aerospace Sci.*, October, 741–87.

Roulin, J. (1833) Théorie de la terre d'après M. Ampère. *Revue des Deux Mondes*, **3**, 96–107.

Slichting, H. (1968) *Boundary Layer Theory*. McGraw-Hill.

Sloudsky, T. H. (1895) De la rotation de la terre supposée fluide à son intérieur. *Bull. Soc. Imp. Nat. Moscou*, **IX**, 285–318.

Smylie, D. E. (1974) Dynamics of the outer core. *Veroff. Zentralinst. Phys. Erde. Berlin*, **30**, 91–104.

Smylie, D. E. and Rochester, M. G. (1981) Compressibility, core dynamics and the subseismic wave equation. *Phys. Earth Plan. Int.*, **24**, 308–19.

Spiegel, E. A. and Veronis, G. (1960) On the Boussinesq approximation for a compressible fluid. *Astr. J.* **131**, 442–7.

Stergiopoulos, S. and Aldridge, K. D. (1982) Inertial waves in a fluid partially filling a cylindrical cavity during spin-up from rest. *Geophys. Astr. Fluid Dyn.*, **21**, 89–112.

Stergiopoulos, S. and Aldridge, K. D. (1984) Ringdown of inertial waves in a spheroidal shell of rotating fluid. *Phys. Earth Plan. Int.*, **36**, 17–26.

Stern, M. E. (1963) Trapping of low frequency oscillations in an equatorial boundary layer. *Tellus*, **15**, 246–50.

Stewartson, K. (1953) On the flow between two rotating coaxial disks. *Proc. Camb. Phil. Soc.*, **49**, 333–41.

Stewartson, K. (1957) On almost rigid rotation. *J. Fluid Mech.*, **3**, 17–26.

Stewartson, K. (1960a) Motion of bodies through conducting fluids. *Rev. Mod. Phys.* **32**(4), 855–7.

Stewartson, K. (1960b) On the motion of a non-conducting body through a perfectly conducting fluid. *J. Fluid Mech.*, **8**, 82–96.

Stewartson, K. and Rickard, J. R. A. (1969) Pathological oscillations of a rotating fluid. *J. Fluid Mech.*, **35**, 759–73.

Stewartson, K. and Roberts, P. H. (1963) On the motion of a liquid in a spheroidal cavity of a precessing rigid body. I. *J. Fluid Mech.*, **17**, 1–20.

Stewartson, K. and Ward, G. N. (1958) On the stability of a spinning top containing liquid. *J. Fluid Mech.*, **5**, 577–92.

Taylor, G. I. (1923) Stability of a viscous liquid contained between two rotating cylinders. *Phil. Trans. Roy. Soc. Lond.* **A223**, 289.

Taylor, G. I. (1936) Fluid friction between rotating cylinders. I. Torque measurements. *Proc. Roy. Soc. Lond.*, **A157**, 546.

Thomson W. (Kelvin) (1880) On gravitational oscillations of rotating water. *Phil. Mag.* **X**, 109–16.

Tillman, W. (1961) Zum Reibungsmoment der Turbulenten Stroemung Zwischen Rotierenden Zylindern. *Forsch. Ing.-Wes.*, **27**, 189.

Todoeschuck, J. P. and Rochester, M. G. (1980) The effect of compressible flow on anti-dynamo theorems. *Nature*, **284**, 250–1.

Tolstoy, I. (1963) The theory of waves on stratified fluids including the effects of gravity and rotation. *Rev. Mod. Phys.*, **35**, 207–30.

Toomre, A. (1966) On the coupling of the Earth's core and mantle during the 26000 year precession. *The Earth–Moon System*, Plenum Press, 33–45.

Vanyo, J. P. (1974) Transformation from a 7-dimensional experimental space for precessing fluid energy to a 2-parameter analytical space. *J. Appl. Mech.* (*Brief Notes*), 1128–30.

Vanyo, J. P. and Likins, P. W. (1972) Rigid-body approximations to turbulent motion in a liquid-filled, precessing, spherical cavity. *J. Appl. Mech.*, 18–23.

Vanyo, J. P. and Paltridge, G. W. (1981) A model for energy dissipation at the mantle–core boundary. *Geophys. J. R. Astr. Soc.*, **66**, 677–90.

Vanyo, J. P., Lu, V. C. and Weyant, T. F. (1975) Dimensionless energy dissipation for precessional flows in the region of $Re = 1$. *J. Appl. Mech.* (*Brief Notes*), p. 881.

Venezian, G. (1966) Flow in a precessing spherical enclosure. 2: Low-viscosity. Rep. 85–34, Div. Eng. Appl. Sci. Calif. Inst. Technol.

Veronis, G. (1968) Large-amplitude Benard convection in a rotating fluid. *J. Fluid Mech.*, **31**(1), 113–39.

Veronis, G. (1970) The analogy between rotating and stratified fluids. *Ann. Rev. Fluid. Mech.*, **2**, 37–67.

Wedemeyer, E. H. (1964) The unsteady flow within a spinning cylinder. *J. Fluid Mech.*, **20**(3), 383–99.

Weidman, P. D. (1976) On the spin-up and spin-down of a rotating fluid. Part 1: extending the Wedemeyer model; Part 2: Measurements and stability. *J. Fluid Mech.*, **77**, 685–708; 709–36.

Wood, W. W. (1966) An oscillatory disturbance of rigidly rotating fluid. *Proc. Roy. Soc. Lond.* **A293**, 181–212.

CHAPTER 4

Geomagnetism

4.1 Fundamental laws of electromagnetism

These laws are the mathematical expressions of a number of experimental facts. A fluid is electrically conducting when it contains free charges and the phenomenon of magnetism is produced when electrically charged particles are in motion: it is convenient to describe such an effect by introducing *the magnetic field vector* $\bar{B}$ (magnetic induction).

One also defines $\bar{H}$, the intensity of the magnetic field with

$$\bar{B} = \mu \bar{H} \tag{4.1}$$

μ being called "permeability" and uniformly taken as equal to the permeability of free space $\mu_0 = 10^{-6}$ henry m^{-1}.

The Gauss law

The experimental fact is that isolated magnetic charges have never been observed and consequently cannot exist upon a *closed* surface; also magnetic field lines can never end, though they do not, in general, form closed loops. This can be expressed by the equation of flow Φ:

$$\Phi = \oiint_S \bar{B} \cdot \hat{n}\, dS = 0 \tag{4.2}$$

or

$$\iiint_V \operatorname{div} \bar{B}\, dV = 0 \tag{4.3}$$

(Green–Ostrogradsky theorem). Thus, in the volume limited by S, we have

$$\operatorname{div} \bar{B} = 0 \tag{4.4}$$

so that longitudinal electromagnetic waves do not exist, only shear transverse waves.

One can now define a new vector $\bar{A}$, called *magnetic vector potential*, such that:

$$\bar{B} = \operatorname{curl} \bar{A} \tag{4.5}$$

while

$$\oiint_S \operatorname{curl} \bar{A}\cdot\hat{n}\,dS = \oint_C \bar{A}\cdot dx \tag{4.6}$$

But

$$\bar{A}' = \bar{A} + \operatorname{grad}\psi \tag{4.7}$$

is also a convenient solution and this arbitrariness can be used to require $\bar{A}$ to satisfy an additional condition that we can choose to simplify the formalism. It looks convenient to put:

$$\operatorname{div}\bar{A} = 0 \tag{4.8}$$

The Ampère and Kirchhoff laws

The *circulation* (or work) of the magnetic field vector $\bar{B}$ along a closed contour C is proportional to the flux of the electric current density $\bar{J}$ across any surface S supported by this contour:

$$\oint_C \bar{B}\cdot d\bar{x} = \oiint_S \operatorname{curl}\bar{B}\cdot\hat{n}\,dS = \mu\oiint_S \bar{J}\cdot\hat{n}\,dS \tag{4.9}$$

It results from this that

$$\begin{aligned} \operatorname{curl}\bar{B} &= \mu\bar{J} \quad \text{(Ampère)} \\ \operatorname{div}\bar{J} &= 0 \quad\ \ \text{(Kirchhoff)} \end{aligned} \tag{4.10}$$

If the loop C is a plane loop in the plane OXY, $\hat{n}$ has the same direction as Oz and J has only one component J_z. If the loop is tilted with respect to the plane OXY, $\bar{J}$ has three components. We can write

$$\operatorname{curl}\operatorname{curl}\bar{A} = \operatorname{grad}\operatorname{div}\bar{A} - \operatorname{div}\operatorname{grad}\bar{A} = \mu\bar{J} \tag{4.11}$$

and as we have chosen A in such a way that $\operatorname{div} A = 0$, we find a Poisson equation:

$$\nabla^2\bar{A} = -\mu\bar{J} \tag{4.12}$$

which, in gaussian units, is written $\nabla^2\bar{A} = -4\pi\mu\bar{J}$ to be compared with the Poisson equation

$$\nabla^2 V = -4\pi G\rho$$

$\bar{J}$ is clearly a vector density of current (amp m^{-2})

The Lenz–Faraday law or induction law (1831)

If the magnetic field is changing with time, relative to a certain frame of reference, then a particle will undergo, per unit of its charge, a force called the *induced electric field* E_i. From

$$E_i = \oint_C \bar{E}_i \, d\bar{x} = -\frac{d\Phi}{dt} \tag{4.13}$$

we have

$$\oiint_S \text{curl } \bar{E}_i \cdot \hat{n} \, dS = -\oiint_S \frac{\partial \bar{B}}{\partial t} \cdot \hat{n} \, dS \tag{4.14}$$

from which

$$\text{curl } \bar{E}_i = -\frac{\partial \bar{B}}{\partial t} \tag{4.15}$$

In electrostatics one simply has

$$\text{curl } \bar{E}_S = 0 \qquad \bar{E}_S = -\text{grad } \phi \tag{4.16}$$

ϕ being the Coulomb electrostatic potential. Here we obtain from $\bar{B} = \text{curl } \bar{A}$:

$$\bar{E}_i = -\frac{\partial \bar{A}}{\partial t} \tag{4.17}$$

The minus sign means that the direction of the induced electromagnetic force is opposed to the change which generates it.

The Ohm law

In electrostatics this law is expressed by the equation

$$\bar{J} = \sigma \bar{E}_S \tag{4.18}$$

where

$$\bar{E}_S = -\text{grad } \phi \tag{4.19}$$

is the electrostatic field and σ the electric conductivity.

In electromagnetism we observe a *motional induction effect* due to the motion of velocity $\bar{q}$ so that

$$\bar{J} = \sigma(\bar{E} + \bar{q} \wedge \bar{B}) \tag{4.20}$$

this equation describes the effect of fluid motion on the electromagnetic field.

We will introduce later the Lorentz force, which describes the inverse effect of the magnetic field upon the fluid motion and has, therefore, to be introduced in the Navier–Stokes equation.

We have thus obtained the four Maxwell equations

$$\left.\begin{aligned}
&\text{curl } \bar{B} = \mu \bar{J} && \text{(Ampère)} \\
&\text{div } \bar{B} = 0 && \text{(Gauss)} \\
&\text{curl } \bar{E} = -\frac{\partial \bar{B}}{\partial t} && \text{(Lenz–Faraday)} \\
&\bar{J} = \sigma(\bar{E} + \bar{q} \wedge \bar{B}) && \text{(Ohm)}
\end{aligned}\right\} \tag{4.21}$$

which describe experimental facts in the most convenient mathematical form. As $\bar{j}$ and $\bar{E}$ can be deduced from $\bar{B}$ and $\bar{q}$, the magnetohydrodynamics will essentially concern the study of interactions between the $\bar{q}$ and $\bar{B}$ fields through the induction and Navier equations.

Note: We are not concerned here with the propagation of the electromagnetic waves; we can thus ignore the displacement currents in Maxwell's equations.

Combining these equations give

$$\frac{\partial \bar{B}}{\partial t} = -\operatorname{curl}\left(\frac{1}{\sigma}\bar{J} - \bar{q}\wedge\bar{B}\right) \tag{4.22}$$

or

$$\frac{\partial \bar{B}}{\partial t} = \underbrace{-\operatorname{curl}\left(\frac{1}{\mu\sigma}\operatorname{curl}\bar{B}\right)}_{\text{ohmic dissipation}} + \underbrace{\operatorname{curl}(\bar{q}\wedge\bar{B})}_{\text{motional induction}} \tag{4.23}$$

Putting $\eta^* = (\mu\sigma)^{-1}$ we can write it under this final form:

$$\frac{\partial \bar{B}}{\partial t} = \operatorname{curl}(\bar{q}\wedge\bar{B}) + \eta^*\nabla^2\bar{B} = \bar{B}\operatorname{grad}\bar{q} - \bar{q}\operatorname{grad}\bar{B} + \eta^*\nabla^2\bar{B} \tag{4.24}$$

which is the magnetic induction equation. We see that $\eta^*\nabla^2\bar{B}$ is a diffusive process and η^* a "*magnetic viscosity*".

If there were no convective motion ($\bar{q} = 0$) the field would simply decay with time according to the law:

$$\frac{\partial \bar{B}}{\partial t} = \eta^*\nabla^2\bar{B} \tag{4.25}$$

with a time constant: $\tau = L^2/\eta^*$ called Cowling constant. With $\sigma = 3 \times 10^5\ \mathrm{S\,m^{-1}}$, $\mu = 10^{-6}\ \mathrm{H\,m^{-1}}$ one has $\eta^* \sim 3\times 10^4\ \mathrm{cm^2\,s^{-1}}$, which means that it should take only some 10,000 years for a magnetic field to disappear from the Earth's core (with $L \cong 10^8$ cm). However paleomagnetism has demonstrated that the age of the geomagnetic field is more than 2×10^9 years.

Thus some source of mechanical energy must be found to maintain the field with $\bar{q} \neq 0$: the energy budget will be evaluated in §4.11. Unfortunately we have no direct knowledge of $\bar{q}$ and $\bar{B}$ while the indirect knowledge leaves much to be desired.

We can, anyway, state that an initial magnetic field $\bar{B}$ and a fluid velocity field $\bar{q}$ induce an electrical field $\bar{q}\wedge\bar{B}$. If the medium is conducting, a volume current $\bar{J}$ will be driven by this electrical field with an additional magnetic field associated to it. If the latter reinforces the original field, we can have a *self-excited dynamo* and the field no longer needs to be maintained by external sources: the energy lost through ohmic dissipation is compensated by conversion of mechanical into magnetic energy. Equation (4.24) also shows that a toroidal fluid motion $\bar{q}$ does not interact with a toroidal magnetic field: the $\bar{q}$ field must correspond to movements *across* the lines of force of the magnetic field.

4.2 Main features of the geomagnetic field

If one agrees that most of the earth's mantle is an *insulating* shell, $\bar{j} = 0$ in it, thus $\nabla \wedge \bar{B} = 0$ and $\bar{B} = -\nabla V$. As moreover $\nabla \cdot \bar{B} = 0$ the magnetic potential V is harmonic outside the core and one may write (Gauss, 1839):

$$V = a \sum_{n=1}^{N} \sum_{m=0}^{n} \left(\frac{a}{r}\right)^{n+1} [g_n^m(t) \cos m\lambda + h_n^m(t) \sin m\lambda]\, P_n^m(\theta) \quad r < a \tag{4.26}$$

for $r > r_{\text{core}}$, a being the mean radius of the Earth.

This field has no toroidal part as $\bar{B}$ derives from a potential and consequently its lines of force cannot be closed lines, being orthogonal to the equipotentials. The field we describe in this way is thus a poloidal field. However, in the nearly perfectly conducting core, there is no reason why one should not have a more or less intense toroidal field. But we see here that such a field cannot be observed at the surface of the Earth so that we can only make conjectures about its intensity. Different models are thus proposed for the geodynamo, depending upon the hypothesis made about the intensity of the toroidal part of the field.

The radial component of the magnetic field $\bar{B} = -\nabla V$ is given by

$$B_r(r, \theta, \lambda, t) = \sum_{n=1}^{N} (n+1) \left(\frac{a}{r}\right)^{n+2}$$

$$\sum_{m=0}^{n} [g_n^m(t) \cos m\lambda + h_n^m(t) \sin m\lambda]\, P_n^m(\theta) \tag{4.27}$$

The list of g_n^m, h_n^m coefficients is called the *Definitive International Geomagnetic Reference Field* (DGRF). It is given for a specified epoch t and is now complete up to the order $n = 10$, $m = 10$ (table 4.1). The terms of order $n = 1$ represent a centered "dipole" for which the *surface* observations give, in nanotesla:

$$\left.\begin{array}{llll} \text{for 1965} & g_1^0 = -30334\,\gamma & g_1^1 = -2119\,\gamma & h_1^1 = +5776\,\gamma^* \\ \text{for 1980} & g_1^0 = -29988\,\gamma & g_1^1 = -1957\,\gamma & h_1^1 = +5606\,\gamma \end{array}\right\} \tag{4.28}$$

that is about 0.35 gauss at the equator, which corresponds to about 4 gauss at

* 1 tesla = 1 weber m^{-2} = 10^4 gauss; 1 gamma = 10^{-5} gauss = 1 nanotesla (nT)

TABLE 4.1 *Spherical harmonic coefficients g_n^m and h_n^m (in nT) for International Geomagnetic Reference Field (IGRF) 1980, and Definitive International Geomagnetic Reference Field (DGRF) 1965–75*

		DGRF (1965)		*DGRF* (1970)		*DGRF* (1975)		*IGRF* (1980)		1980–85 in nT/yr	
n	m	g_n^m	h_n^m	g_n^m	h_n^m	g_n^m	h_n^m	g_n^m	h_n^m	g_n^m	h_n^m
1	0	−30334		−30220		−30100		−29988		22.4	
1	1	−2119	5776	−2068	5737	−2013	5675	−1957	5606	11.3	−15.9
2	0	−1662		−1781		−1902		−1997		−18.3	
2	1	2997	−2016	3000	−2047	3010	−2067	3028	−2129	3.2	−12.7
2	2	1594	114	1611	25	1632	−68	1662	−199	7.0	−25.2
3	0	1297		1287		1276		1279		0.0	
3	1	−2038	−404	−2091	−366	−2144	−333	−2181	−335	−6.5	0.2
3	2	1292	240	1278	251	1260	262	1251	271	−0.7	2.7
3	3	856	−165	838	−196	830	−223	833	−252	1.0	−7.9
4	0	957		952		946		938		−1.4	
4	1	804	148	800	167	791	191	783	212	−1.4	4.6
4	2	479	−269	461	−266	438	−265	398	−257	−8.2	1.6
4	3	−390	13	−395	26	−405	39	−419	53	−1.8	2.9
4	4	252	−269	234	−279	216	−288	199	−298	−5.0	0.4
5	0	−219		−216		−218		−219		1.5	
5	1	358	19	359	26	356	31	357	46	0.4	1.8
5	2	254	128	262	139	264	148	261	149	−0.8	−0.4
5	3	−31	−126	−42	−139	−59	−152	−74	−150	−3.3	0.0
5	4	−157	−97	−160	−91	−159	−83	−162	−78	0.2	1.3
5	5	−62	81	−56	83	−49	88	−48	92	1.4	2.1
6	0	45		43		45		49		0.4	
6	1	61	−11	64	−12	66	−13	65	−15	0.0	−0.5
6	2	8	100	15	100	28	99	42	93	3.4	−1.4
6	3	−228	68	−212	72	−198	75	−192	71	0.8	0.0
6	4	4	−32	2	−37	1	−41	4	−43	0.8	−1.6
6	5	1	−8	3	−6	6	−4	14	−2	0.3	0.5
6	6	−111	−7	−112	1	−111	11	−108	17	−0.1	0.0
7	0	75		72		71		70		−1.0	
7	1	−57	−61	−57	−70	−56	−77	−59	−83	−0.8	−0.4
7	2	4	−27	1	−27	1	−26	2	−28	0.4	0.4
7	3	13	−2	14	−4	16	−5	20	−5	0.5	0.2
7	4	−26	6	−22	8	−14	10	−13	16	1.6	1.4
7	5	−6	26	−2	23	0	22	1	18	0.1	−0.5
7	6	13	−23	13	−23	12	−23	11	−23	0.1	−0.1
7	7	1	−12	−2	−11	−5	−12	−2	−10	0.0	1.1
8	0	13		14		14		20		0.8	
8	1	5	7	6	7	6	6	7	7	−0.2	−0.1
8	2	−4	−12	−2	−15	−1	−16	1	−18	−0.3	−0.7
8	3	−14	9	−13	6	−12	4	−11	4	0.3	0.0
8	4	0	−16	−3	−17	−8	−19	−7	−22	−0.8	−0.8
8	5	8	4	5	6	4	6	4	9	−0.2	0.2
8	6	−1	24	0	21	0	18	3	16	0.7	0.2
8	7	11	−3	11	−6	10	−10	7	−13	−0.3	−1.1
8	8	4	−17	3	−16	1	−17	−1	−15	1.2	0.8
9	0	8		8		7		6			
9	1	10	−22	10	−21	10	−21	11	−21		
9	2	2	15	2	16	2	16	2	16		
9	3	−13	7	−12	6	−12	7	−12	9		
9	4	10	−4	10	−4	10	−4	9	−5		

TABLE 4.1—(*continued*)

		DGRF (1965)		*DGRF* (1970)		*DGRF* (1975)		*IGRF* (1980)		1980–85 *in nT/yr*	
n	m	g_n^m	h_n^m	g_n^m	h_n^m	g_n^m	h_n^m	g_n^m	h_n^m	g_n^m	h_n^m
9	5	−1	−5	−1	−5	−1	−5	−3	−7		
9	6	−1	10	0	10	−1	10	−1	9		
9	7	5	10	3	11	4	11	7	10		
9	8	1	−4	1	−2	1	−3	1	−6		
9	9	−2	1	−1	1	−2	1	−5	2		
10	0	−2		−3		−3		−3			
10	1	−3	2	−3	1	−3	1	−4	1		
10	2	2	1	2	1	2	1	2	1		
10	3	−5	2	−5	3	−5	3	−5	2		
10	4	−2	6	−1	4	−2	4	−2	5		
10	5	4	−4	6	−4	5	−4	5	−4		
10	6	4	0	4	0	4	−1	3	−1		
10	7	0	−2	1	−1	1	−1	1	−2		
10	8	2	3	0	3	0	3	2	4		
10	9	2	0	3	1	3	1	3	−1		
10	10	0	−6	−1	−4	−1	−5	0	−6		

It is interesting to point out that the order 6 represents practically the totality of the field. In the core of radius 3485 km the order 6 corresponds to about 600 km wave length.

the surface of the core† while its angular position is given by

$$\left.\begin{array}{lll} \text{for 1965} & \tan\lambda_0 = \dfrac{h_1^1}{g_1^1} = -2.726 \rightarrow \lambda_0 = 110^\circ 15' & \\ \text{for 1980} & \qquad\quad = -2.865 \qquad\quad 109^\circ 14' & \\ \text{for 1965} & \operatorname{cotan}\theta_0 = g_1^0/\sqrt{(g_1^1)^2 + (h_1^1)^2} = -4{,}930 \rightarrow \theta_0 = 168^\circ 33' & \\ \text{for 1980} & \qquad\qquad -5{,}050 \qquad\quad 168^\circ 48' & \end{array}\right\} \quad (4.29)$$

The north magnetic pole is in the southern hemisphere at 11°27′ from the geographic south pole. It is not known why this is so.

Other spherical harmonic models exist up to higher degree-and-order, based on ground and airborne data (WC 80, up to $n, m = 12$), on satellite, airborne and ground data (U 061380, up to $n, m = 24$) (GSFC 1266, up to $n, m = 10$), on magnetically quiet days of MAGSAT data (MGST 680, up to $n, m = 13$).

The Faraday concept of lines of force gives an aesthetic picture on how the field B varies in the earth's surrounding space. As shown on figure 4.1, these lines of force are oriented from the positive pole, in our southern hemisphere, towards the negative pole, in our northern hemisphere. The spacing between two lines being inversely proportional to the intensity of the field, their bundle is more tightened in the vicinity of the poles (where B is about 0.7 gauss as

† for $n = 1: (n+1)(a/r)^{n+2} = 2(a/r)^3 = 12.2$ at the core boundary.

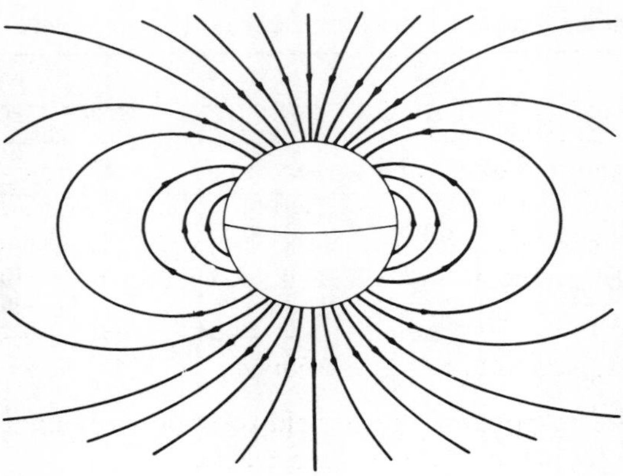

FIGURE 4.1 Lines of force of the geomagnetic field are oriented from the southern geographical hemisphere (north magnetic pole) to the northern geographical hemisphere (south magnetic pole).

$P_n^m(\theta = 0) = 1$) than around the equator (where B is about 0.35 gauss as $P_n^m(\theta = 90°) = 1/2$).

The lines of force are drawn in such a way that the number of lines per unit normal cross-section area is proportional to the field intensity.

At present the moment of the dipole (4.28) decreases by 0.08% per year while it precesses towards the west 0.05° per year in longitude (about 0.2 mm s^{-1}) and 0.02° per year in latitude.

The *non-dipole field* drifts westward 0.2° per year, which corresponds to 0.7 mm s^{-1} at the core equator. Its moment varies by about 10 nT per year.

Inversions which are still unexplained, have happened several hundred times since precambrian times (10^9 years) without any apparent regularity.

Geomagnetic maps, like one shown on the figure 4.2, look similar to meteorological maps and may suggest that there is turbulence inside the core. The eddies responsible for the observed patterns are of continental size. The sources responsible for such features cannot, therefore, be very deep inside the core but are most probably situated in its external layers.

The speeds of fluid motions in the core, that we can infer from these observed drifts of geomagnetic field patterns and intensity (0.03 to 4 cm s^{-1}), are thus many orders of magnitude lower than the speeds of sound waves (seismic waves: $\sim$ 8 km s^{-1}). This is why, when dealing with geomagnetic phenomena, *we will treat the core as an incompressible fluid*, that is:

$$\nabla \cdot \bar{q} = 0 \tag{4.30}$$

The observed drifts could be ascribed either to hydromagnetic wave propagations inside the core (see 4.7) either to a differential rotation of the core with respect to the mantle, which should rotate faster than the core if a *westward* drift is to be explained in this way. This hypothesis is supported by the correlation with the variations of the Earth's rotation (figure 1.7).

If, as proposed in chapter 1, the inner core is progressively created by iron precipitation from the eutectic mixture, it implies that the convection has a radial component and the conservation of momentum requires that the inner layers rotate more rapidly than the outer layers. Then the westward drift could be interpreted on the result of the slower rotation of the upper layers. Such a differential rotation will obviously generate toroidal fields.

It is not to be excluded that both processes are acting, but it is at present impossible to determine how much each contributes.

Downward extrapolation to the core boundary

The downward extrapolation of the field to the core boundary is obtained multiplying the DGRF coefficients by a factor

$$(n+1)\left(\frac{a}{r}\right)^{n+2} \quad \text{with} \quad \frac{a}{r} = \frac{6371}{3485} = 1.828 \tag{4.31}$$

which is 12.2 for $n = 1$ and dramatically increases with n. The procedure considerably amplifies the higher harmonics which are the less well determined, and therefore simultaneously amplifies the quite large errors at these high orders.

An example of such an extrapolation is given on the figure 4.2, taken from Booker (1969).

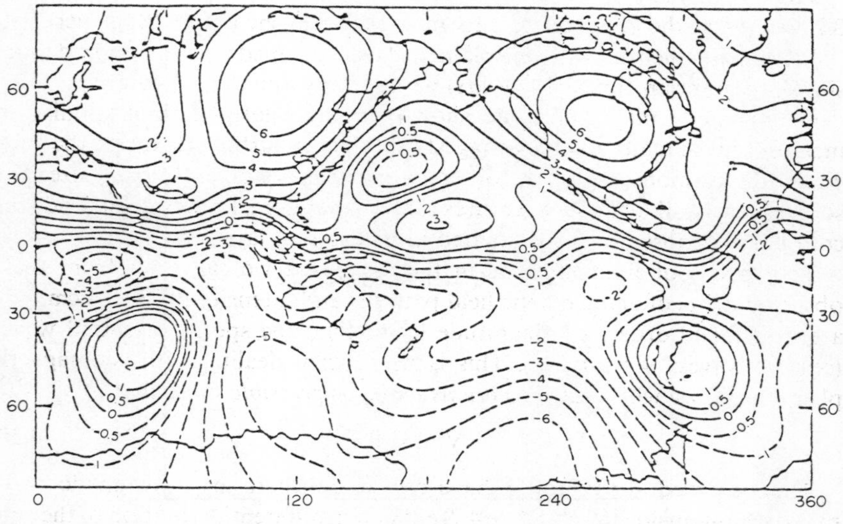

FIGURE 4.2 Mercator projection of B_r on the core ($r_c = 3500$ km) evaluated with model GSFC (12/66) truncated above degree 7. The contour interval is 1 G except that the ± 0.5 G contours are included in some regions to clarify the structure (Booker, 1969).

The power spectrum of the field produced by harmonics of a given order n (wavelength $2\pi r/n$), on a surface of radius r, is defined (Booker, 1969; Lowes, 1966, 1974) by

$$R_n = (n+1) \sum_{m=0}^{n} [(g_n^m)^2 + (h_n^m)^2] \cdot \left(\frac{a}{r}\right)^{2n+4} \tag{4.32}$$

and is expressed in nT^2 (10^{-10} gauss^2).

Putting R_n in a graph in function of n shows a characteristic change of slope at about $n = 14$ which Langel and Estes (1982) explain by attributing to the field in the core the terms with orders n up to $n \leq 13$ and to the crustal fields the terms with orders $n \geq 15$ (see figure 4.3).

Langel and Estes use MAGSAT data and have proposed the following adjustments by linear regressions:

from $n =$ 2 to 12: $R_n = 1.349 \times 10^9 \; (0.270)^n (nT)^2$ for the core (4.33)

from $n =$ 16 to 23: $R_n = 37.1 \; (0.974)^n (nT)^2$ for the crust (4.34)

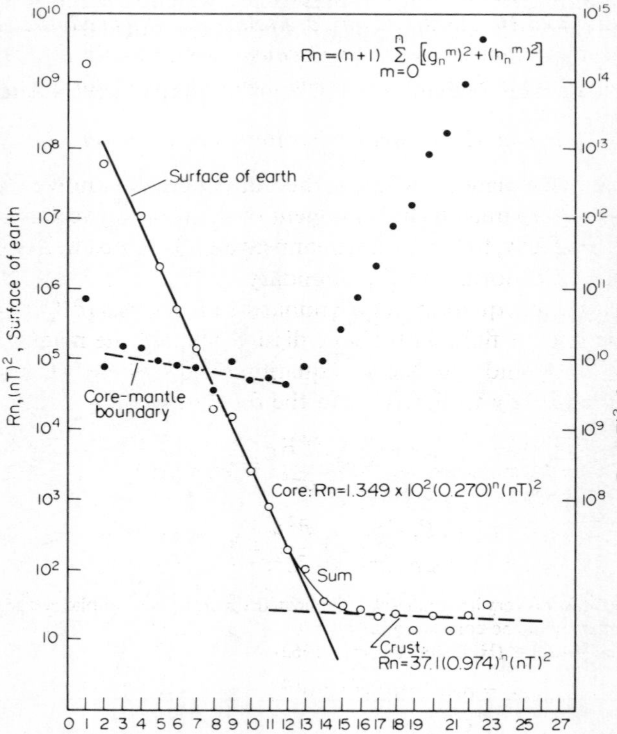

FIGURE 4.3 Geomagnetic field spectrum at the surface of the Earth and at the surface of core (downward extrapolation) according to Langel and Estes (1982).

We can summarize the observed features of the geomagnetic field as follows

Dipole field

1. its intensity has decreased by about 0.08 % per year between 1965 and 1980;
2. it precesses westwards by 0.05°/year;
3. it drifts northwards by about 2 km/year or 0.02°/year.

Non-dipole field

1. it represents about 20 % of the dipole field;
2. it precesses westwards by about 0.2°/year;
3. its intensity fluctuates by about 10 γ/year.

Observed periodicities

One has observed "short" periods of 11, 20, 60 years and "long" periods of 360, 600 years and perhaps 900 and 1200 years. This is the reason why we can here treat the fluid in the core as incompressible (but see §3.2): sound waves have speeds of several km per second, which is incompatible with these observed periods. Such periods are still short with respect to the viscous and ohmic decay times of the irreversible processes, and to treat these fluctuations one can consider them as adiabatic.

Acceleration impulses (jerks) of the geomagnetic field have been observed, according to Le Mouel and Courtillot (1981) around 1840, 1905 and 1969 with time constants less than 1 year *and in concomitance with changes in the Earth's rotation behaviour* (figure 1.7; see also Kerridge and Barraclough, 1985).

At such a short time scale the core must behave as a perfect conductor so that the lines of force of the magnetic field remain "frozen" in the fluid.

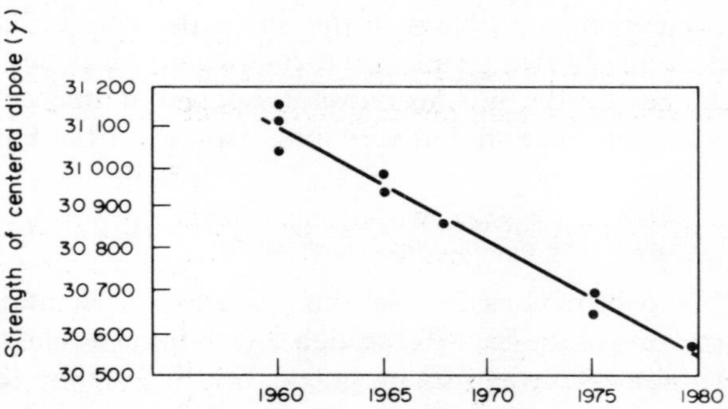

FIGURE 4.4 Field decay

Inversions

1. the polarity of the field is inverted with a frequency which is quite variable: 4 or 5 times in 1 million years, but for periods of 10 million years there are sometimes no inversions;
2. the duration of this process is also difficult to determine: an estimation of 10,000 years has been proposed but it may be very different;
3. the field intensity decreases abruptly at the time of the inversions.

Magsat models

These show that the magnetic flux leaves and re-enters the core in a small number of areas or spots of great intensity.

Dipole representation

A representation with a convenient number of dipoles situated inside the core can also fit the surface observations. A loop of radius r where a current of intensity i circulates has a magnetic moment

$$\begin{aligned} p_m &= \pi r^2 i = \pi r^2 \frac{q}{T} = \pi r^2 q \frac{v}{2\pi r} = \frac{1}{2} q v r \\ &= \frac{1}{2} \frac{q}{m} (m v r) = \frac{1}{2} \frac{q}{m} L \end{aligned} \tag{4.35}$$

where L is the angular momentum.

However the calculation is not linear, and therefore more complicated than the spherical harmonics procedure. As the field is known only at the surface of the Earth this representation is not unique, and hypotheses have to be introduced concerning the depth of the dipoles.

Remark

There is no comparison possible with the Sun's magnetic field because the geodynamo is deeply buried inside a non-conductive container, the mantle, which is not the case for the Sun. Moreover, the observed time variations have very short periods on the Sun, but very long periods on the Earth.

A correlation between the geomagnetic and gravity fields

A result which may be of considerable consequences for the interpretation of the internal structure of the Earth is the high correlation found by Hide (1970) and Hide and Malin (1970) between the gravity field and the geomagnetic field: a shift in longitude of the geomagnetic field raises the correlation coefficient up to a significant 0.84 value when the displacement is 160° for 1965. This displacement has indeed to be a function of time because of the non-dipole westward drift. Hide and Malin propose

$$\Delta\lambda = (126.2 \pm 0.2)^\circ + (0.273 \pm 0.005)^\circ\ (t - 1835 \pm 10) \tag{4.36}$$

where the annual variation 0.27° corresponds to this drift. This corresponds to a zero shift some 500 years ago.

4.3 Penetration of the magnetic field and currents into the lower mantle: skin depth

To simplify the matter let us suppose that we have, in the core, a field

$$\vec{B}(B_0 e^{-i\omega t}, 0, 0) \tag{4.37}$$

Therefore, in the conducting lower mantle, we should have from eqn. 4.24

$$\left.\begin{aligned} &\eta^* \frac{\partial^2 B_x}{\partial z^2} = \frac{\partial B_x}{\partial t} = -i\omega B_x \\ &B_y = B_z = 0 \\ &E_x = E_z = 0 \quad (\bar{E} = \eta^* \operatorname{curl} \bar{B}) \end{aligned}\right\} \tag{4.38}$$

The solution is of the wave form

$$B_x = B_0 e^{\lambda z - i\omega t} \tag{4.39}$$

with

$$\eta^* \lambda^2 = -i\omega = \omega e^{-i\pi/2} \tag{4.40}$$

or

$$\lambda = \pm \left(\frac{\omega}{\eta^*}\right)^{1/2} e^{-i\pi/4} = \pm (1 - i)\left(\frac{\omega}{2\eta^*}\right)^{1/2} \tag{4.41}$$

the plus sign is to be rejected as it produces an exponential increase of the field. Thus

$$\bar{B}\ (\text{mantle}) \equiv B_0\,(\text{core}) e^{-z\sqrt{\omega/2\eta^*}} \exp i\left[z\left(\frac{\omega}{2\eta^*}\right)^{1/2} - \omega t\right], 0, 0 \tag{4.42}$$

The distance at which the amplitude is reduced by a factor 1/e is called the

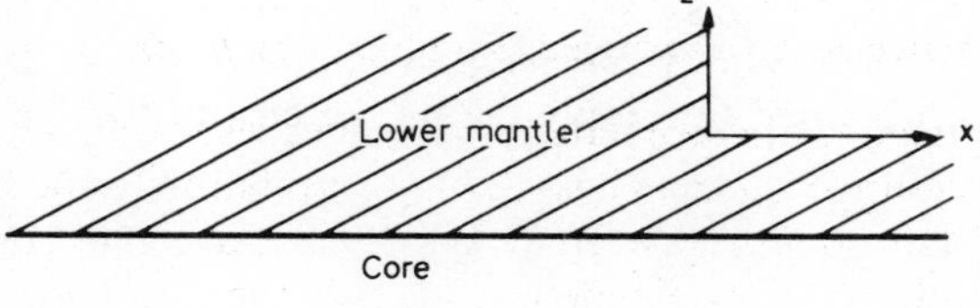

FIGURE 4.5.

electromagnetic skin depth. It is:

$$d = \left(\frac{2\eta^*}{\omega}\right)^{1/2} = (\eta^* T)^{1/2}\left(\frac{1}{\pi}\right)^{1/2} = \frac{\sqrt{2}}{\sqrt{\mu\sigma\omega}} \tag{4.43}$$

Taking $\sigma = 3 \times 10^5\ \mathrm{S\,m^{-1}}$ (see eqn 2.38) one obtains from eqn (4.43), in the core:

	$d =$ 16 metres	for $T =$	4 minutes
	58		54 minutes
	300		1 day
	1640		30 days
	5733		1 year
and	1812 km	for $T =$	100 000 years

It results that a layer of perfect conductivity ($\sigma = \infty$) acts as a shield for a *variable* magnetic field: $d = 0$ from eqn 4.43.
The characteristic time $\eta^{*-1}d^2 = \mu\sigma d^2$ is called the Cowling constant.

Thus fast waves (short period oscillations) do not penetrate very far but induce eddy currents in the lower mantle. This creates a form of electromagnetic coupling between the core and the mantle.

A serious problem is the uncertainty of our knowledge of the electrical conductivity of the lower mantle.

Anyway short period fluctuations (periods of few years) in the core are screened by the skin effect in the mantle and cannot be observed at the Earth's surface; they can nevertheless couple the mantle to the core in its rotation.

But the fact that variations of the magnetic field at time scales of decades are observed after having penetrated the whole mantle means that the global mantle electric conductivity is small.

According to Currie (1967) magnetic signals generated in the core with periods less than about 4 years are cut off by the mantle acting as a low pass filter.

4.4 Analogy between the magnetic induction equation and the vorticity equation (Batchelor, 1950)

It is instructive to show how the induction equation

$$\frac{\partial \bar{B}}{\partial t} = \nabla\Lambda(\bar{q}\Lambda\bar{B}) + (\eta^*\nabla^2\bar{B}) \qquad \nabla\cdot\bar{B} = 0 \tag{4.44}$$

looks similar to the *vorticity Helmholtz equation* (3.60) of hydrodynamics which has been obtained by simply taking the curl of the Navier equation:

$$\frac{\partial \bar{\omega}}{\partial t} = \nabla\Lambda(\bar{q}\Lambda\bar{\omega}) + (\nu\nabla^2\bar{\omega}) \qquad \nabla\cdot\bar{\omega} = 0 \tag{4.45}$$

(the gradients evidently disappear through this operation). Let us forget about the dissipative terms ($\nu\nabla^2\bar{\omega}$) and transform (4.45) by introducing the material derivative:

$$\frac{D\bar{\omega}}{Dt} = (\bar{\omega}\cdot\text{grad})\,\bar{q} \tag{4.46}$$

where we shall take, for example, the z component:

$$\omega_z \frac{\partial w}{\partial z} = (\bar{\omega}\cdot\nabla\bar{q})_z \tag{4.47}$$

if $\partial w/\partial z > 0$ we will observe an elongation of the vortex in the z direction. To conserve mass in an incompressible fluid there must be a contraction in the x, y directions and to conserve momentum it must rotate faster: eqn (4.47) shows how, in that case, the *vorticity increases.* This is called *vortex stretching.* In a similar way it is possible that $\bar{B}$ increases under corresponding conditions. However, to trigger the geodynamo, a small initial magnetic field is needed which can be a result of the weak field which permeates the galaxy. We must pay attention however that $\bar{\omega}$ is related to $\bar{q}$ (by $\bar{\omega} = \text{curl}\ \bar{q}$) while $\bar{B}$ is not. Moreover the boundary conditions will be totally different.

Other analogies between geomagnetism and continuum mechanics are given later with the definitions of Maxwell magnetic pressure, shear stress and magnetic viscosity.

The "frozen field"

If there is no dissipation, the vortex lines move with the fluid as well as the magnetic lines of force. One usually says that they are "frozen" in the fluid. They can be considered as "tracers" for the flow and the geomagnetic westward drift appears as the image of the slower rotation of the upper layers of the core.

4.5 Magnetic dimensionless characteristic numbers

Reynolds and Prandtl magnetic numbers

The analogy, just emphasized, very logically induces us to introduce magnetic analogues to the hydrodynamic dimensionless numbers. We therefore have

a Reynolds hydrodynamic number $R = \dfrac{LU}{\nu}$ (3 parameters) (4.48)

a Reynolds magnetic number $R_m = \dfrac{LU}{\eta^*}$ (3 parameters) (4.49)

a Prandtl thermal number $P = \dfrac{\nu}{\chi}$ (2 parameters) (4.50)

a Prandtl magnetic number $P_m = \dfrac{\nu}{\eta^*}$ (2 parameters) (4.51)

In principle when $R_m \gg 1$ the magnetic diffusion is weak and the situation is close to a "*frozen flux*". One has, however, to take care that the length scales can be much larger in the $\bar{B}$ equation than in the $\bar{q}$ equation.

On the other hand, if P_m is low, as in metallic liquids and in the Earth's core ($\sim 10^{-6}$), the magnetic field diffuses more quickly than vorticity, and one could expect that the magnetic boundary layer would be thicker than the viscous boundary layer (but see Ekman–Hartmann composite boundary layer (§4.10)).

We can see here why the realization of a model in the laboratory is practically impossible.

The conductivity of the fluids which are at our disposal (for example Hg, liquid Na) is not great enough to obtain a large value for R_m which can only reach about unity.

At the same time, however, P_m is extremely small (1.5×10^{-7} for Hg) so that the Reynolds hydrodynamic number $R = R_m/P_m$ becomes enormous, which means that there is turbulence which makes the observations very imprecise.

In the face of such experimental impossibilities we have no other method than to proceed to the very difficult mathematical developments.

The dimensions of the cosmic bodies are so large that even poorly conducting fluids have a high R_m so that the flows are strongly coupled to the magnetic field.

Reynolds thermal number or Péclet number

The heat transport equation being

$$\frac{\partial T}{\partial t} + \vec{q}\cdot\nabla T = \chi\nabla^2 T + \frac{A}{C_p} \tag{4.52}$$

where A is the density of radiogenic sources, one finds it an advantage to introduce a new dimensional number

$$\text{Pé} = \frac{LU}{\chi} = R\cdot P \quad \text{(3 parameters)} \tag{4.53}$$

This is called the Péclet number, or the Reynolds thermal number.

The three Reynolds numbers allow one to decide if convection processes prevail over the corresponding diffusive processes.

4.6 Lorentz force and Maxwell stresses

In a fluid conductor of electricity, the motions produce a current of density $\vec{J}$ = electrical charge $\times\ \vec{q}$ so that the magnetic field exerts a counteracting force called the *Lorentz force*:

$$\bar{L} = \bar{j}\Lambda\mu\bar{H} = \bar{j}\Lambda\bar{B} \tag{4.54}$$

This force must, of course, be introduced in the budget of forces of the Navier–Stokes equation. Matter, moving with the velocity $\bar{q}$, works against this force:

$$N = -\iiint \bar{q}\cdot\bar{L}\,dV = -\iiint \bar{q}\cdot(\bar{j}\Lambda\mu\bar{H})\,dV \tag{4.55}$$

which has, as a result, to limit the growing of the amplitude of the velocity field $\bar{q}$. The expression of the Lorentz force can be transformed as follows:

$$\bar{j}\Lambda\mu\bar{H} = \text{curl}\,\bar{H}\Lambda\mu\bar{H} = -\mu\bar{H}\,\Lambda\,\text{curl}\,\bar{H} \tag{4.56}$$

where we can introduce

$$\frac{1}{2}\,\overline{\text{grad}}\,H^2 = \bar{H}\Lambda\,\text{curl}\,\bar{H} + \bar{H}\cdot\text{grad}\,\bar{H} \tag{4.57}$$

which gives

$$\bar{j}\Lambda\mu\bar{H} = -\overline{\text{grad}}\,\frac{\mu H^2}{2} + \mu\bar{H}\cdot\overline{\text{grad}}\,\bar{H} \tag{4.58}$$

or

$$\bar{j}\Lambda\bar{B} = \bar{B}\cdot\text{grad}\,\frac{\bar{B}}{\mu} - \text{grad}\,\frac{|\bar{B}|^2}{2\mu} \tag{4.59}$$

on the other hand

$$\text{div}\,\overline{\mu H H} = \mu\bar{H}\,\text{div}\,\bar{H} + \mu\bar{H}\cdot\text{grad}\,\bar{H} \tag{4.60}$$

where

$$\text{div}\,\bar{H} = 0 \tag{4.61}$$

so that, finally:

$$\bar{L} = \bar{j}\Lambda\mu\bar{H} = -\overline{\text{grad}}\,\frac{\mu H^2}{2} + \text{div}\,\mu\bar{H}\cdot\bar{H} \tag{4.62}$$

$\mu H^2/2$ has the dimension of a pressure. We call it the *isotropic Maxwell magnetic pressure.* It is to be introduced within the reduced pressure, just like the centrifugal potential so that we shall put, in the conservation Navier equation:

$$p = \frac{1}{\rho}P - \frac{1}{2}|(\bar{\Omega}\Lambda\bar{r})|^2 + \frac{\mu H^2}{2\rho} \tag{4.63}$$

It may be interesting to develop (4.59) in more detail taking for example, its x component:

$$(\bar{B}\cdot\overline{\text{grad}})\frac{B_x}{\mu} - \frac{\partial}{\partial x}\frac{|\bar{B}|^2}{\mu^2} = \frac{\partial}{\partial x}\left(B_x^2 - \frac{|\bar{B}|^2}{2}\right)\frac{1}{\mu} + \frac{\partial}{\partial y}\left(\frac{B_x B_y}{\mu}\right) + \frac{\partial}{\partial z}\left(\frac{B_x B_z}{\mu}\right) \tag{4.64}$$

which is obvious as div $\overline{B}=0$, and to compare it to the expression of an elastic stress per unit of volume in x direction:

$$\frac{\partial \tau_{xx}}{\partial x}+\frac{\partial \tau_{xy}}{\partial y}+\frac{\partial \tau_{xz}}{\partial z} \tag{4.65}$$

By analogy, one can put here:

$$\left.\begin{aligned}
&\tau_{xx}=\left(B_x^2-\frac{|\overline{B}|^2}{2}\right)\frac{1}{\mu}, && \tau_{xy}=\frac{B_xB_y}{\mu}, && \tau_{xz}=\frac{B_xB_z}{\mu}\\
&\tau_{yx}=\frac{B_yB_x}{\mu} && \tau_{yy}=\left(B_y^2-\frac{|\overline{B}|^2}{2}\right)\frac{1}{\mu}, && \tau_{yz}=\frac{B_yB_z}{\mu}\\
&\tau_{zx}=\frac{B_zB_y}{\mu} && \tau_{zy}=\frac{B_zB_y}{\mu} && \tau_{zz}=\left(B_z^2-\frac{|\overline{B}|^2}{2}\right)\frac{1}{\mu}
\end{aligned}\right\} \tag{4.66}$$

The term

$$\pi=\frac{|\overline{B}|^2}{2\mu}=\frac{\mu H^2}{2} \tag{4.67}$$

is indeed the "*isotropic Maxwell pressure*" or "*magnetic pressure*".

The components τ_{xy}, τ_{xz}, . . . are the *Maxwell magnetic shear stresses.* They allow a liquid to support shear waves. The Maxwell pressure increases the pressure across the magnetic lines but not along them. This destroys hydrostatic equilibrium. It results that the velocity of sound waves $V_0=\sqrt{k/\rho}$ depends upon their azimuth θ with respect to the field direction:

$$V=\frac{1}{2}(V_0^2+a^2+2V_0a\cos\theta)^{1/2}+\frac{1}{2}(V_0^2+a^2-2V_0a\cos\theta)^{1/2}$$

Note: There is also an analogy between the expression (4.59) of the Lorentz force and the advection term in (3.40):

$$(\bar{q}\cdot\nabla)\bar{q}=\text{curl }\bar{q}\Lambda\bar{q}+\text{grad}\,\frac{|q|^2}{2}$$

or

$$\bar{\xi}\Lambda\bar{q}=(\bar{q}\cdot\text{grad})\,\bar{q}-\text{grad}\,\frac{|q|^2}{2}$$

4.7 Magnetohydrodynamic waves: the Alfvén waves

One can imagine that, when the core–mantle boundary is excited by seismic waves, the fluid in the core is slightly perturbed and magnetohydrodynamic waves are stimulated.

The divergence term in the Lorentz force represents a tensile stress acting as a restoring force on the lines of force, resulting in shear waves (because div $\bar{B} = 0$) which propagate along these lines of force as in the case of an elastic harp string when it is stretched by external forces (this is a kind of pseudoelasticity of the fluid). Such transverse waves are not possible in a liquid unless a magnetic field permeates it.

By analogy with the vibrating strings we can thus write the velocity as

$$a = \sqrt{\frac{\text{tension}}{\text{linear density}}} = \sqrt{\frac{\mu H^2}{\rho}} = \frac{B}{\sqrt{\mu\rho}} \tag{4.68}$$

Such waves, discovered by Alfvén, are called *Alfvén waves* and a is the Alfvén speed. It does not depend upon the wave length: such waves are non-dispersive. However if we introduce the effects of Earth's rotation, a polarization appears and the waves become dispersive.

We see that, with $\mu = 10^{-6}\ H\,\text{m}^{-1}$ and $\rho = 10^4\ \text{kg m}^{-3}$, a field of 4 gauss gives $a = 0.4\ \text{cm s}^{-1}$ (i.e. travelling 2500 km within 20 years—remember that a longitudinal seismic P wave travels across the core in 20 min.) compatible with the observations of the poloidal geomagnetic non-dipole field. A 100 gauss toroidal field could generate Alfvén waves with a speed of $10\ \text{cm s}^{-1}$ (much more than the west-ward drift 1 gauss $\rightarrow$ 1 mm s^{-1}).

In a fluid with ohmic and viscous dissipations we consider a small perturbation $\bar{h}$ of a uniform magnetic field $\bar{H}$; neglecting square and rectangular terms in $\bar{h}$ and $\bar{q}$, we have then the simultaneous equations:

$$\left.\begin{aligned}
\frac{\partial \bar{q}}{\partial t} &= \frac{1}{\rho}\bar{B}\cdot\nabla\bar{h} - \nabla p + \nu\nabla^2\bar{q} \\
\frac{\partial \bar{h}}{\partial t} &= \bar{H}\cdot\nabla\bar{q} + \eta^*\nabla^2\bar{h} \\
p &= \frac{P}{\rho} + \frac{\mu}{\rho}\bar{H}\cdot\bar{h} \\
\nabla\cdot\bar{q} = 0 &\qquad \nabla\cdot\bar{h} = 0
\end{aligned}\right\} \tag{4.69}$$

(the only non-negligible term in curl $(\bar{q}\wedge(\bar{H}+\bar{h})$ is indeed $\bar{H}\cdot\text{grad}\ \bar{q}$). When applying the divergence operator to the first of these equations we obtain

$$\nabla^2 p = 0 \tag{4.70}$$

and $p = 0$ is a solution relevant to the problem on hand.

Seeking for a plane wave solution with a space–time dependency for $\bar{q}$ and $\bar{h}$ in exp $i\ (\bar{k}\cdot\bar{r} - \omega t)$ gives:

$$\left.\begin{aligned}
-i\omega\bar{q} &= \frac{i\mu}{\rho}(\bar{k}\cdot\bar{H})\bar{h} - k^2\nu\bar{q} \\
-i\omega\bar{h} &= i(\bar{k}\cdot\bar{H})\bar{q} - k^2\eta^*\bar{h}
\end{aligned}\right\} \tag{4.71}$$

or

$$\left.\begin{aligned}(\omega+ik^2\nu)\bar{q} &= -\frac{\mu}{\rho}(\bar{k}\cdot\bar{H})\bar{h}\\(\omega+ik^2\eta^*)\bar{h} &= -(\bar{k}\cdot\bar{H})\bar{q}\end{aligned}\right\} \tag{4.72}$$

from which

$$(\omega+ik^2\nu)(\omega+ik^2\eta^*) = \frac{\mu}{\rho}(\bar{k}\cdot\bar{H})^2 = \frac{\mu}{\rho}(H\cos\theta)^2k^2 \tag{4.73}$$

or, with $a = (\sqrt{\mu H^2/\rho})\cos\theta$

$$\{i\omega(\eta^*+\nu)-a^2\}k^2-\eta^*\nu k^4+\omega^2 = 0 \tag{4.74}$$

if the wave vector makes an angle θ with the magnetic field vector.

For a perfect fluid without ohmic dissipation ($\nu = \eta^* = 0$):

$$\omega = \sqrt{\frac{\mu}{\rho}}Hk\cos\theta = \frac{B\cos\theta}{\sqrt{\mu\rho}}k = ak \tag{4.75}$$

$$\frac{\partial\omega}{\partial k} = \pm a$$

these waves are not dispersive (linear orbits in planes parallel to the wave front).

In the case of a perfect fluid with ohmic dissipation

$$\omega(\omega+ik^2\eta^*) = a^2k^2 \tag{4.76}$$

is the *dispersion relation*: a depends upon the wave length $\lambda = 2\pi/k$.

Note: Without dissipations and taking $\bar{H}$ parallel to Oz, eqn (4.69) can be written simply

$$\frac{\partial\bar{h}}{\partial t} = \bar{H}\frac{\partial\bar{q}}{\partial z}, \qquad \frac{\partial\bar{q}}{\partial t} = \frac{\mu H}{\rho}\frac{\partial\bar{h}}{\partial z} \tag{4.77}$$

from which

$$\frac{\partial^2\bar{h}}{\partial t^2} = a^2\frac{\partial^2\bar{h}}{\partial z^2}, \quad \frac{\partial^2\bar{q}}{\partial t^2} = a^2\frac{\partial^2\bar{q}}{\partial z^2} \tag{4.78}$$

hyperbolic equations having waveform solutions with the Alfvén velocity

$$a = \frac{\mu H}{\sqrt{\mu\rho}} = \frac{B}{\sqrt{\mu\rho}} \tag{4.79}$$

4.8 Magnetostrophic waves.
(see Moffat, 1970; Acheson and Hide, 1973)

The Lorentz force is obviously the only force which could balance the Coriolis force in the core, which has an order of magnitude:

$$2\bar{q}\Lambda\bar{\Omega} \therefore 2\Omega U \sim 1.5\times 10^{-8}\ \mathrm{m\,s^{-2}} \quad \text{(per mass unit)} \tag{4.80}$$

One should thus need that

$$\frac{1}{\mu}\bar{B}\cdot\nabla\bar{B} \therefore \frac{B^2}{\mu\rho L}\sim 10^{-8}\ \mathrm{m\,s^{-2}} \quad \text{(per mass unit)} \tag{4.81}$$

which implies a field B of about 200 gauss. This would be called a "*magnetostrophic equilibrium*" defined by the condition

$$B\approx\sqrt{2\Omega\mu\rho LU}. \tag{4.82}$$

As it is known from observations that the poloidal field reaches only 4 gauss, the only hypothesis we could make in this sense is to assign such a strength to the toroidal field, which is what some authors, e.g. Braginsky, indeed do.

We call magnetostrophic those inertial hydromagnetic waves which are still controlled by the balance of the Coriolis and Lorentz forces. We consider here the usual *inviscid incompressible rotating* fluid pervaded by a *uniform* magnetic field $\bar{H}$ with a small fluctuation field $\bar{h}$ induced by the motion $\bar{q}$ across $\bar{H}$. The equations are:

$$\left.\begin{aligned}&\frac{\partial\bar{q}}{\partial t}+2\bar{\Omega}\Lambda\bar{q}=-\nabla p+\mu\frac{\bar{H}}{\rho}\cdot\nabla\bar{h}\\ &\frac{\partial\bar{h}}{\partial t}=\bar{H}\cdot\nabla\bar{q}+\eta^{*}\nabla^2\bar{h}\\ &\nabla\cdot\bar{q}=0\qquad\nabla\cdot\bar{h}=0\end{aligned}\right\} \tag{4.83}$$

$$p=\frac{P}{\rho}-\frac{1}{2}(\bar{\Omega}\Lambda\bar{r})^2+(2\mu\rho)^{-1}B^2 \tag{4.84}$$

TABLE 4.2 *Order of magnitude of the different forces*

with $\Omega = 7\times10^{-5}$ rad s^{-1} $L\sim10^{8}$ cm
$q\sim10^{-2}$ cm s^{-1} $\rho\sim10$
$B\sim200$ gauss $\nu\sim10^{-2}$ stokes unit: cm s^{-2}
one should have

	$\partial\bar{q}/\partial t$	10^{-12}?
Advection	$(\bar{q}\cdot\nabla)\bar{q}$	10^{-12}
Coriolis	$2\bar{\Omega}\Lambda\bar{q}$	max: 1.5×10^{-6}
Viscous	$\nu\nabla^2\bar{q}$	10^{-20}
Lorentz	$\frac{1}{\mu}\bar{B}\cdot\nabla\bar{B}$	4×10^{-5}

but the Coriolis force is zero at the equator! and no longer controls the horizontal flow.

being the reduced pressure, and we seek again waveform solutions with space–time dependence $\exp i(\bar{k}\cdot\bar{r}-\omega t)$. It results

$$-i\omega\bar{q}+2\bar{\Omega}\Lambda\bar{q} = -i\bar{k}p+i(\bar{H}\cdot\bar{k})\bar{h}\frac{\mu}{\rho} \tag{4.85}$$

$$-i\omega\bar{h} = i(\bar{H}\cdot\bar{k})\bar{q}-\eta^* k^2\bar{h} \tag{4.86}$$

$$\bar{k}\cdot\bar{q} = 0 \qquad \bar{k}\cdot\bar{h} = 0 \tag{4.87}$$

the wave vector $\bar{k}$ is orthogonal to $\bar{h}$ as well as to $\bar{q}$: such waves are transverse waves.

Equation (4.86) gives an important relation between $\bar{h}$ and $\bar{q}$

$$\bar{h} = -\frac{\bar{H}\cdot\bar{k}}{\omega+i\eta^* k^2}\bar{q} \tag{4.88}$$

and then eqn (4.85) is written:

$$i\sigma\bar{q}+2\bar{\Omega}\Lambda\bar{q} = -i\bar{k}p \tag{4.89}$$

with

$$\sigma = -\omega+(\bar{H}\cdot\bar{k})^2(\omega+i\eta^* k^2)^{-1}\frac{\mu}{\rho} \tag{4.90}$$

multiplying (4.89) vectorially two times by $\bar{k}$ gives successively:

$$i\sigma\bar{k}\Lambda\bar{q}-2(\bar{k}\cdot\bar{\Omega})\bar{q} = 0$$

$$i\sigma k^2\bar{q}+2(\bar{k}\cdot\bar{\Omega})\bar{k}\Lambda\bar{q} = 0$$

$$i\sigma k^2\bar{q}+2(\bar{k}\cdot\bar{\Omega})\frac{2(\bar{k}\cdot\bar{\Omega})\bar{q}}{i\sigma} = 0$$

$$\sigma^2 k^2 = 4(\bar{k}\cdot\bar{\Omega})^2$$

and finally

$$\sigma = \pm\frac{2(\bar{k}\cdot\bar{\Omega})}{k} \tag{4.91}$$

and

$$\xi = i\bar{k}\Lambda\bar{q} = \frac{2(\bar{k}\cdot\bar{\Omega})}{\sigma}\bar{q} = \pm k\bar{q} \tag{4.92}$$

Non-dissipative case

We consider now eqn (4.83) without magnetic dissipation ($\eta^* = 0$), derive the first one again with respect to t:

$$\frac{\partial^2\bar{q}}{\partial t^2}+2\bar{\Omega}\Lambda\frac{\partial\bar{q}}{\partial t} = \frac{\mu}{\rho}\bar{H}\cdot\nabla\frac{\partial\bar{h}}{\partial t}-\nabla\frac{\partial p}{\partial t} \tag{4.93}$$

and replace $\partial \bar{h}/\partial t$ by $\bar{H}\cdot\nabla\bar{q}$. Then we take the curl and obtain

$$\left(\frac{\partial^2}{\partial t^2}-(a\cdot\nabla)^2\right)\operatorname{curl}\bar{q}-(2\bar{\Omega}\cdot\nabla)\frac{\partial\bar{q}}{\partial t}=0 \tag{4.94}$$

$\left(\text{as curl}\left(2\bar{\Omega}\wedge\frac{\partial\bar{q}}{\partial t}\right)=2\bar{\Omega}\cdot\text{grad}\,\frac{\partial\bar{q}}{\partial t}\right).$

We can seek for plane wave solutions by introducing the usual form $\exp i(\bar{k}\cdot\bar{r}-\omega t)$ into (4.94) which immediately gives the dispersion relation:

$$\omega^2+(2\bar{\Omega}\cdot\bar{k})\frac{\omega}{k}-(\bar{a}\cdot\bar{k})^2=0 \tag{4.95}$$

one can simply extract the roots and take the square:

$$\omega^2=(\bar{a}\cdot\bar{k})^2+\frac{1}{2}\left\{\frac{(2\bar{\Omega}\cdot\bar{k})^2}{k^2}\pm\sqrt{\frac{(2\bar{\Omega}\cdot\bar{k})^4}{k^4}+4\,\frac{(\bar{a}\cdot\bar{k})^2\,(2\bar{\Omega}\cdot\bar{k})^2}{k^2}}\right\} \tag{4.96}$$

or

$$\left.\begin{aligned}\omega^2&=\omega_m^2+\frac{1}{2}\{\omega_r^2\pm(\omega_r^4+4\omega_m^2\omega_r^2)^{1/2}\}^*\\ \omega^2&=(\bar{a}\cdot\bar{k})^2+\frac{1}{2}\frac{(2\bar{\Omega}\cdot\bar{k})^2}{k^2}\left\{1\pm\sqrt{1+\frac{4(\bar{a}\cdot\bar{k})^2k^2}{(2\bar{\Omega}\cdot\bar{k})^2}}\right\}\end{aligned}\right\} \tag{4.97}$$

The square root of $(1+s)$ is $1+\frac{s}{2}-\frac{s^2}{8}+\ldots$, thus,

if

$$\frac{(\bar{a}\cdot\bar{k})^2\,k^2}{(2\bar{\Omega}\cdot\bar{k})^2}\ll 1 \tag{4.98}$$

$$\omega_-^2=\frac{(\bar{a}\cdot\bar{k})^4k^2}{(2\bar{\Omega}\cdot\bar{k})^2}\quad\text{hydromagnetic wave, “slow mode”} \tag{4.99}$$

while

$$\omega_+^2=\frac{(2\bar{\Omega}\cdot\bar{k})^2}{k^2}\quad\text{“fast mode”} \tag{4.100}$$

and this is the frequency of the purely inertial wave (as a does not enter into account).

Magnetic modes are slower than inertial modes. On the contrary if

$$\frac{(\bar{a}\cdot\bar{k})^2k^2}{(2\bar{\Omega}\cdot\bar{k})^2}\gg 1 \tag{4.101}$$

* Hide and Stewartson, 1972.

one can simplify (4.97) as follows

$$\begin{aligned}\omega_+^2 &\simeq (\bar{a}\cdot\bar{k})^2\left\{1+\left|\frac{2\bar{\Omega}\cdot\bar{k}}{k(\bar{a}\cdot\bar{k})}\right|\right\}\\ \omega_-^2 &\simeq (\bar{a}\cdot\bar{k})^2\left\{1-\left|\frac{2\bar{\Omega}\cdot\bar{k}}{k(\bar{a}\cdot\bar{k})}\right|\right\}\end{aligned} \tag{4.102}$$

showing a small frequency splitting due to the rotation Ω.

One has evidently

$$\omega_-^2 \leq (\bar{a}\cdot\bar{k})^2 \leq \omega_+^2 \tag{4.103}$$

the equality corresponding to $\Omega = 0$.

One also has:

$$\omega_+^2\,\omega_-^2 = \omega_m^4 \tag{4.104}$$

All these waves are dispersive.

4.9 The Hartmann number

In 1937 J. Hartmann stated that "the effect of the magnetic field on a laminar flow is to increase the apparent viscosity approximately proportionally to the field intensity". It is easy to see that it is indeed so.

The Lorentz force $\bar{j}\Lambda\mu\bar{H}$ being perpendicular to $\bar{H}$, it does not act along the lines of force but transversal (t) to these lines. Thus we may write:

$$\bar{j}\Lambda\mu\bar{H} = \mu\sigma(\bar{E}+\mu\bar{q}\Lambda\bar{H})\Lambda\bar{H} = \mu\sigma(\bar{E}_t\Lambda\bar{H}-\mu H^2\bar{q}_t) \tag{4.105}$$

so that a simplified form of the Navier equation becomes:

$$\rho\,(d\bar{q}/dt) = \rho\nabla p + \mu\sigma(\bar{E}_t\Lambda\bar{H} - \mu H^2\bar{q}_t) \tag{4.106}$$

and, if ∇p and $\bar{E}_t$ were negligible:

$$(d\bar{q}_t/dt) + \left(\frac{\sigma\mu^2}{\rho}H^2\right)\bar{q}_t = 0 \tag{4.107}$$

an equation which represents an oscillation with a damping time

$$\tau = \rho\,(\sigma\mu^2H^2)^{-1} \tag{4.108}$$

Here the Lorentz force $(\bar{j}\Lambda\bar{H})$ plays the role of a *magnetic viscosity* and if V is used as symbol of speed:

the magnetic viscous force per volume unit is $\sigma\mu^2H^2V$

the dynamic viscous force $(\rho\nu\nabla^2\bar{q})$ is $\rho\nu VL^{-2}$

The square root of their ratio

$$M = \mu HL\,(\sigma/\rho\nu)^{1/2} \tag{4.109}$$

is called the *Hartmann number*, which gives a measure of the capacity of magnetic viscosity to dominate hydrodynamic viscosity.

One also often introduces as dimensionless number, the *Prandtl magnetic number*:

$$P_m = \frac{\nu}{\eta^*} \tag{4.110}$$

For the Earth's core

$$P_m \cong \frac{4 \times 10^{-6}\ \mathrm{m^2\,s^{-1}}}{2.6\ \mathrm{m^2\,s^{-1}}} \sim 1{,}5 \times 10^{-6} \tag{4.111}$$

Thus ohmic dissipation exceeds viscous dissipation by several orders of magnitude.

The Hartmann number is large in laboratory experiments, which simplifies the demonstration of the effect of the magnetic forces upon fluid motions but inversely makes it practically impossible to demonstrate experimentally the effect of fluid motions on the magnetic field.

4.10 Hydromagnetic boundary layers

What we call a "boundary" is in fact a discontinuity, so that the equations of mechanics cannot be resolved in such a region. As mentioned before, they must be replaced by an *equal* number of discontinuity conditions across the boundary: each equation corresponds to a number of conditions, this number being equal to the order of the differential equation. Important consequences follow from this:

1. For a viscous fluid the Navier differential equation is one order higher than for a perfect fluid, so that one boundary condition has to be added. As a discontinuity in the tangential velocity (q_s) would imply an infinite gradient of velocity and an infinite shear stress, it was natural to introduce the condition:

$$q_s = 0 \tag{4.112}$$

2. When we add a magnetic field this does not increase the order of the differential equation with respect to the order of viscous Navier–Stokes equations. We will thus have to combine Newtonian viscosity and magnetic viscosity into one *unique* additional boundary condition; this is possible by combining the Maxwell stresses with the viscous stresses.
3. Geostrophic flow corresponds to equations of lower order (by 1). It is thus no more possible to satisfy all boundary conditions, and this implies the existence of restricted areas with strong gradients (fronts).

Speaking in terms of viscosity and combining together laminar, turbulent and magnetic viscosity within the $\nu\nabla^2\bar{q}$ term of the equations, it becomes clear that we will have only one additional boundary condition and one composite boundary layer where the different viscosities act according to their relative importance.

The Ekman–Hartmann composite boundary layer

We consider here a fluid in contact with a solid along an interface S. Stewartson (1957) demonstrated that if the normal component of the magnetic field across S is non-zero, a discontinuity in the tangential components along S (to which correspond surface currents) would be dispersed instantaneously as an Alfvén wave in the fluid. One can thus conclude that, if the normal component is non-zero, the tangential components of the magnetic field must be continuous at the boundary of a perfectly conducting fluid.

Let us consider now a fluid core with viscous coefficient ν and magnetic viscosity:

$$\eta^* = (\mu\sigma)^{-1}. \tag{4.113}$$

where σ is the conductivity and

$$R_m = \frac{UL}{\eta^*} \tag{4.114}$$

the Reynolds magnetic number.

We may assume that $\nu = \eta^* = 0$ in the bulk of the core but not inside the boundary layer(s). This has as a consequence to lower the order of the governing differential equations by 2, and it therefore becomes impossible for the main stream to satisfy the no-slip conditions for the velocity $\bar{q}$:

$$\bar{q} = 0 \text{ on } S \tag{4.115}$$

and the conditions

$$\bar{B} \text{ continuous across } S \tag{4.116}$$

The role of the boundary layer is precisely to provide the necessary adjustments between the bulk of the core and the interface.

In hydrodynamics we only put $\nu = 0$ (perfect fluid) but this also reduces the order by 2, we can satisfy only one condition, namely

$$q_r = 0 \tag{4.117}$$

and the solution will not satisfy the conditions of no-slip

$$q_\theta = q_\lambda = 0 \tag{4.118}$$

This situation was corrected by a boundary layer (Ekman layer in rotating fluids) of thickness $(\nu\tau)^{1/2}$ in which the dissipating diffusive term $\nu\nabla^2$ is crucial.

This boundary layer matches with the solution for the interior of the fluid at its inner edge and with all three continuity conditions (4.115), (4.116) on S.

Similarly, in the absence of fluid motions, putting $\eta^* = 0$ also reduces the order by 2 and the number of boundary conditions must accordingly also be reduced by 2. It is then required to satisfy only the continuity of B_r, and here again a boundary layer of thickness $(\eta^* \tau)^{1/2}$ will satisfy the continuity conditions for B_θ, B_λ on S and match the solution for the interior at its inner edge.

In the Earth's core both phenomena act, fluid motions and magnetic fields, but the simultaneous simplification $\nu = 0$, $\eta^* = 0$ reduces the order by only 2 again, *not by* 4!

The number of boundary conditions must accordingly be reduced by 2 only, and not by 4 as could happen by combining the two preceding arguments. The "intriguing" question could thus be raised as to whether we would keep the conditions on q_θ, q_λ or the conditions on B_θ, B_λ (Roberts and Scott, 1965).

The answer is obtained by combining the viscous stresses with the Maxwell fictitious stresses. Then, as shown by Stewartson, the viscous and electromagnetic diffusive effects are concentrated inside a unique boundary shear layer of thickness δ_H (see also Roberts and Scott, 1965).

This shear layer enables rapid changes to be made in the tangential components of $\bar{q}$ *and* $\bar{H}$. These *changes are related and depend on the relative sizes of* ν *and* η^*, *i.e. on whether the viscous or electrically diffusive forces predominate. This is the Ekman–Hartmann composite boundary layer.*

Thickness of the Ekman–Hartmann composite boundary layer

Let us consider, in the plane xz, a flow in the x direction. We know that, to have an Ekman layer, there must be a component of the rotation vector normal to the boundary. Similarly, to have a Hartmann layer, there must be a component of the magnetic field normal to the boundary: B_z.*

The boundary layer equations for a laminar steady motion ($\partial B/\partial t = 0$) of an incompressible viscous fluid with ohmic dissipation, are the magnetic induction equation (4.24) and the Navier equation (3.40), where we neglect the gradients $\partial/\partial x$ and $\partial/\partial y$ with respect to the $\partial/\partial z$ gradient:

$$
\begin{aligned}
&B_z \frac{\partial u}{\partial z} + \eta^* \frac{\partial^2 B_x}{\partial z^2} = 0 \\
&\frac{B_z}{\mu\rho} \frac{\partial B_x}{\partial z} + \nu \frac{\partial^2 u}{\partial z^2} = 0 \quad \dagger
\end{aligned}
\tag{4.119}
$$

* Such a boundary layer is very irregular and complicated, owing to the complicated pattern of the magnetic field intensity at the core boundary.

† The solution of (4.119) is (Hide and Roberts, 1962):

$$B_z = A + C \exp\{az/(\eta^* \nu)^{1/2}\}$$

$$w = \frac{C}{B_z} a \left(\frac{\eta^*}{\nu}\right)^{1/2} (1 - \exp\{az/(\eta^* \nu)^{1/2}\}) \quad a = B_z/(\mu\rho)^{1/2}$$

w vanishing at $z = 0$ according to the boundary conditions.

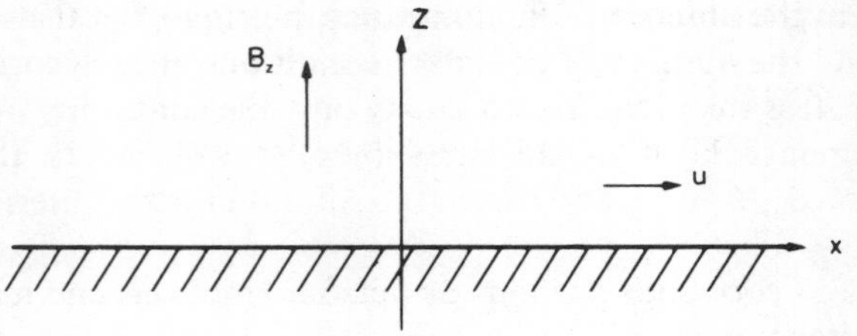

FIGURE 4.6

The first one is the expression of the balance between the fluid flow and the magnetic viscosity, while the second expresses the balance between the Lorentz force and the kinematic viscosity. B_z is assumed constant and w zero inside the boundary layer.

Inside this layer it is convenient to use the stretched coordinate:

$$\xi = \frac{z}{\delta}, \quad \delta_E = \left(\frac{\nu}{\Omega}\right)^{1/2} \tag{4.120}$$

scaled to the boundary layer thickness. Then we write

$$\left.\begin{aligned} B_z \frac{\partial u}{\partial \xi} + \frac{\eta^*}{\delta} \frac{\partial^2 B_x}{\partial \xi^2} &= 0 \\ \frac{B_z}{\mu\rho} \frac{\partial B_x}{\partial \xi} + \frac{\nu}{\delta} \frac{\partial^2 u}{\partial \xi^2} &= 0 \end{aligned}\right\} \tag{4.121}$$

Let us rederive the first one with respect to ξ

$$\frac{\partial}{\partial \xi} \frac{\partial^2 B_x}{\partial \xi^2} + \frac{\delta}{\eta^*} B_z \frac{\partial^2 u}{\partial \xi^2} = 0 \tag{4.122}$$

we take $\partial^2 u/\partial \xi^2$ from the second and we obtain:

$$\frac{\partial}{\partial \xi} \left[\frac{\partial^2}{\partial \xi^2} - \frac{\delta^2 B_z^2}{\mu\rho\nu\eta^*} \right] (B_x, u) = 0 \tag{4.123}$$

because the same operation can be done for u by deriving the second equation.

The thickness of the boundary layer is defined in such a way that the ξ derivative inside this layer is of order of unity.

It results that

$$\frac{\delta^2 B_z^2}{\mu\rho\nu\eta^*} \sim 1 \tag{4.124}$$

or

$$\delta = \delta_H = (\mu\rho\nu\eta^*)^{1/2} B^{-1} \tag{4.125}$$

called Hartmann distance or Hartmann depth. Then we can replace:

$$(\mu\eta^*)^{-1} = \sigma$$

$$B/\sqrt{\mu\rho} = a \quad \text{Alfvén velocity}$$

and obtain

either
$$\delta_H = \frac{\sqrt{\eta^* \nu}}{a} \tag{4.126}$$

or
$$\delta_H = \left(\frac{\rho\nu}{\sigma}\right)^{1/2} B^{-1} \tag{4.127}$$

The Hartmann number being, according to (4.109),

$$M = \mu H L (\sigma/\rho\nu)^{1/2} = BL(\mu\rho\nu\eta^*)^{-1/2} \tag{4.128}$$

we have

$$\delta_H = LM^{-1} \tag{4.129}$$

The boundary layer thickness is proportional to M^{-1}.

A "magnetic interaction parameter" is defined as

$$\alpha = \frac{\delta_E}{\delta_H} = \left(\frac{\nu}{\mu\rho\nu\eta^*\Omega}\right)^{1/2} B \tag{4.130}$$

$$\alpha = (\mu\rho\eta^*\Omega)^{-1/2} B = \left(\frac{\sigma}{\rho\Omega}\right)^{1/2} B = \frac{a}{(\eta^*\Omega)^{1/2}} \tag{4.131}$$

The Ekman–Hartmann composite boundary layer evolves from a pure Ekman boundary layer ($B = 0 \rightarrow \alpha = 0$) to a pure Hartmann boundary layer as the α parameter increases with B.*

In the Earth's core, with

$$\left.\begin{aligned} \Omega &= 7 \times 10^{-5}\ \mathrm{rad\,s^{-1}} \\ \sigma &= 3 \times 10^{5}\ \mathrm{S\,m^{-1}} \\ \mu &= 10^{-6}\ \mathrm{H\,m^{-1}} \\ \eta^{*-1} &= \mu\sigma = 3 \times 10^{-1}\ \mathrm{m^{-2}\,s} \\ (\eta^*\Omega)^{-1/2} &= 0.69\ 10^{2}\ \mathrm{m^{-1}\,s} \\ \text{if } a &= 10^{-2}\ \mathrm{m\,s^{-1}} \end{aligned}\right\} \tag{4.132}$$

one should have

$$\alpha = 0.69 \tag{4.133}$$

* Gilman and Benton (1968) obtain the thickness of the Ekman–Hartmann boundary layer as:

$$[(1+\alpha^4)^{1/2} - \alpha^2]^{1/2} \left(\frac{\nu}{\Omega}\right)^{1/2}$$

which corresponds to a transition region. But this is most uncertain in view of the uncertainty of our knowledge of a.

This parameter squared

$$\alpha^2 = \frac{\sigma B^2}{\rho\Omega} \tag{4.134}$$

represents the ratio between the electromagnetic forces to the Coriolis force, and is also called the "*rotational magnetic force coefficient*". If $\alpha^2 \ll 1$ the flow is not significantly affected by the magnetic forces. "It is the ratio of the speed with which electromagnetic disturbances are transmitted by MAC waves, proportional to a^2/Ω, to that at which they are transmitted by diffusive processes, proportional to η^*" (Hide, 1969).

One also often uses the Lundquist number which is defined as follows

$$K = \sigma LB(\mu\rho^{-1})^{1/2} = \mu\sigma L\frac{B}{\sqrt{\mu\rho}} = \mu\sigma La \tag{4.135}$$

it characterizes the degree of coupling between the magnetic field and the motion.

If one puts here $a = 0$ (no magnetic field) one will obtain again the Ekman solution.

For a rotating body one should write the boundary layer equations under the form:

$$\left.\begin{aligned}
\frac{\partial u}{\partial t} - fv &= \frac{a}{\sqrt{\mu\rho}}\frac{\partial B_x}{\partial z} + v\frac{\partial^2 u}{\partial z^2}\\
\frac{\partial v}{\partial t} + fu &= \frac{a}{\sqrt{\mu\rho}}\frac{\partial B_y}{\partial z} + v\frac{\partial^2 v}{\partial z^2}\\
\frac{\partial B_x}{\partial t} &= a\sqrt{\mu\rho}\frac{\partial u}{\partial z} + \eta^*\frac{\partial^2 B_x}{\partial z^2}\\
\frac{\partial B_y}{\partial t} &= a\sqrt{\mu\rho}\frac{\partial v}{\partial z} + \eta^*\frac{\partial^2 B_y}{\partial z^2}
\end{aligned}\right\} \tag{4.136}$$

keeping again only the derivatives with respect to z.

Putting

$$D \equiv \frac{\partial}{\partial z} = \hat{n}\cdot\nabla \tag{4.137}$$

the elimination of three of the four variables u, v, B_x; B_y will give

$$\left[\left(\frac{\partial}{\partial t} - \eta^* D^2\right)\left(\frac{\partial}{\partial t} - vD^2\right) - a^2 D^2\right]^2 \phi + f^2\left(\frac{\partial}{\partial t} - \eta^* D^2\right)^2 \phi = 0 \tag{4.138}$$

that each one of the variables must satisfy. Plane wave solutions $\phi = \phi_0 \exp(i\omega t + nz)$ give the relation

$$[(i\omega-\eta^*n^2)(i\omega-\nu n^2)-a^2n^2]^2+f^2\,(i\omega-\eta^*n^2)^2=0 \quad \text{(Hide, 1969)} \tag{4.139}$$

With $\omega=0$, $f=0$ one gets the Hartmann result (4.126).
The steady flow case ($\omega=0$) is of special interest as it was used by Rochester (1976) to investigate the dissipative effects of core-mantle coupling on the secular decrease of the obliquity of the earth's orbit. The eqn. (4.139) becomes, in that case

$$\left(\frac{\nu}{f}\right)^2 n^4-2\left(\frac{\nu}{f}\right)\alpha^2n^2+(1+\alpha^4)=0$$

and, if one puts

$$n=-\left(\frac{f}{\nu}\right)^{1/2}(1+\alpha^4)^{1/4}e^{i\vartheta}$$

for the solution, the identifications give

$$e^{2i\vartheta}=(1+\alpha^4)^{-1/2}(\alpha^2\pm i)=\cos 2\vartheta\pm i\sin 2\vartheta$$

$$\cos 2\vartheta=\alpha^2/\sqrt{1+\alpha^4},\ \sin 2\vartheta=1/\sqrt{1+\alpha^4}$$

Thus:

$$\vartheta=\frac{1}{2}\cot^{-1}\alpha^2$$

$$n=-\left(\frac{f}{\nu}\right)^{1/2}(1+\alpha^4)^{1/4}(\cos\vartheta\pm i\sin\vartheta)$$

a result indicated by Acheson and Hide (1973, page 200, eqn. 5.18 and 5.19). The stretched coordinate to be introduced in this case would be

$$\xi=\frac{z}{n}$$

which corresponds to a boundary layer of thickness (see eqn. 3.241):

$$\delta=\left(\frac{f}{\nu}\right)^{-1/2}(1+\alpha^4)^{-1/4}\sec\vartheta$$

or (see eqn. 3.44):

$$\delta=L\left(\frac{E}{2}\right)^{1/2}(1+\alpha^4)^{-1/4}\sec\left(\tfrac{1}{2}\cot^{-1}\alpha\right)$$

Gilman and Benton (1968) have shown that, as the radial fluid motion has to cross the axial magnetic field, it is counteracted by the Maxwell tensions, which results in an inhibition of the Ekman suction by the magnetic field. But the rotational shear flow inside the boundary layer twists the magnetic lines of

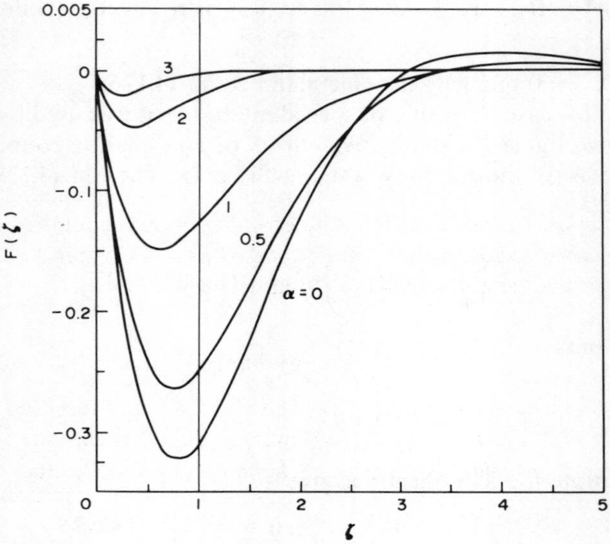

FIGURE 4.7 Radial flow profiles for several values of magnetic interaction parameter.

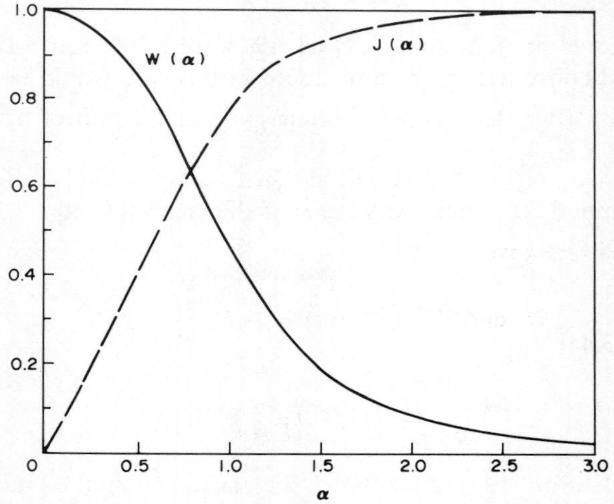

FIGURE 4.8 Normalized Ekman suction velocity and Hartmann current versus magnetic interaction parameter.

forces, and this induces an axial electric current j_z which they called the "Hartmann current".

This current takes the place of Ekman suction and provides a mechanism for spin-up (spin-down) of the core, a spin-up which is more rapid than in the purely hydrodynamic case.

Gilman and Benton (1968) have also shown that the transition from the Ekman to Hartmann limits takes place within a rather small range of α. Gilman (1971) has found that, for a low Prandtl magnetic number, a magnetic field normal to the boundary strongly stabilizes the Ekman layer but that rotation strongly destabilizes the Hartmann layer.

4.11 Energy budget

The evaluation of the energy budget of a geodynamo is an essential step in the search for a mechanism sufficiently powerful to maintain this geodynamo working. It will provide a constraint helping to identify the responsible mechanism.

Magnetic energy

One introduces in electromagnetic theory the concept of the *magnetic energy* accumulated within a body of volume V:

$$W = \frac{1}{2}\iiint_V \bar{H}\cdot\bar{B}\,dV \quad \text{(Reitz and Milford, 1962)} \tag{4.140}$$

We can simply write

$$W = \frac{1}{2\mu}\iiint_V |\bar{B}|^2\,dV \tag{4.141}$$

(it is about 6.10^{22} joules), so that the energy spent by unit of time is

$$\frac{\partial W}{\partial t} = \frac{1}{\mu}\iiint_V \bar{B}\cdot\frac{\partial \bar{B}}{\partial t}\,dV \tag{4.142}$$

Maxwell equations give

$$\bar{H}\cdot\operatorname{curl}\bar{E} - \bar{E}\cdot\operatorname{curl}\bar{H} = -\bar{H}\cdot\frac{\partial \bar{B}}{\partial t} - \bar{E}\cdot\bar{j} \tag{4.143}$$

or

$$\frac{1}{\mu}\bar{B}\cdot\frac{\partial \bar{B}}{\partial t} = -\nabla\cdot(\bar{E}\wedge\bar{H}) - \bar{E}\cdot\bar{j} \tag{4.144}$$

thus

$$\frac{\partial W}{\partial t} = -\oiint_S (\bar{E}\wedge\bar{H})\cdot\hat{n}\,dS - \iiint_V \bar{E}\cdot\bar{j}\,dV \tag{4.145}$$

$(\bar{E} \wedge \bar{H})$ is the vector of Poynting $\bar{S}$. Its flux across the surface S limiting the volume V corresponds to the flux of magnetic energy out of (+) or into (−) the volume, energy which is exchanged with the environment. The advantage of introducing the Poynting vector is that one could calculate this amount on the basis of data observed on the surface S, which in principle would be easier than collecting information everywhere inside the volume V.

If the volume was chosen in such a way that the Poynting flux were zero, this equation should express an energy conservation.

Now, if we use Ohm's law written as

$$\bar{E} = \frac{1}{\sigma}\bar{j} - (\bar{q} \wedge \bar{B}) \tag{4.146}$$

from which

$$\bar{E} \cdot \bar{j} = \frac{1}{\sigma} |\bar{j}|^2 - \bar{j} \cdot (\bar{q} \wedge \bar{B}) = \frac{1}{\sigma} |\bar{j}|^2 - \bar{q} \cdot (\bar{j} \wedge \bar{B}) \tag{4.147}$$

we get the "Poynting theorem":

$$\frac{\partial W}{\partial t} = -\oiint_S (\bar{E} \wedge \bar{H}) \cdot \hat{n} dS - \frac{1}{\sigma} \iiint_V |\bar{j}|^2 dV + \iiint_V \bar{q} \cdot (\bar{j} \wedge \bar{B}) dV \tag{4.148}$$

equation of conservation of the magnetic energy in the volume V.

The last two terms represent transformation of magnetic energy into thermal energy and kinetic energy respectively:

$$Q = \frac{1}{\sigma} \iiint_V |\bar{j}|^2 dV \tag{4.149}$$

is an ohmic dissipation into heat which has *to be added in the heat equation.*

$$L = \iiint_V \bar{q} \cdot (\bar{j} \wedge \bar{B}) dV \tag{4.150}$$

is the kinetic energy created by the Lorentz force $(\bar{j} \wedge \bar{B})$, which is a force of volume and *has to be added in the mechanical energy equation* obtained by multiplication of the Navier equation by $\bar{q}$.

Poynting's theorem is, by itself, a justification of the concept of magnetic energy as it shows that heat production, kinetic energy and flux of the Poynting vector into or out of the closed volume equals the rate of change of W as defined by eqn (4.140). L/Q is the Reynolds magnetic number.

Mechanical energy

With the reduced pressure:

$$p = P - \rho\phi - \tfrac{1}{2}\rho(\bar{\Omega} \wedge \bar{r}) \cdot (\bar{\Omega} \wedge \bar{r}) \tag{4.151}$$

The Navier equation can be written:

$$\frac{D\bar{q}}{Dt}+2\bar{\Omega}\wedge\bar{q}=-\frac{1}{\rho}\nabla p+\frac{1}{\rho}\nabla\cdot\bar{\bar{\tau}} \tag{4.152}$$

If we operate a scalar multiplication by $\bar{q}$, the Coriolis force disappears (we know that it does not "work").

$$\rho\frac{D}{Dt}\left(\frac{q^2}{2}\right)=\bar{q}\cdot(-\nabla p+\nabla\cdot\bar{\bar{\tau}}) \tag{4.153}$$

the three terms representing respectively kinetic, potential and dissipated energy. As

$$-\nabla\cdot(p\bar{q}-\bar{\bar{\tau}}\bar{q})=-\bar{q}\cdot(\nabla p-\nabla\cdot\bar{\bar{\tau}})-(p\nabla\cdot\bar{q}-\bar{\bar{\tau}}\nabla\cdot\bar{q}) \tag{4.154}$$

this conservation equation becomes:

$$\rho\frac{D}{Dt}\left(\frac{q^2}{2}\right)+\nabla\cdot(p\bar{q}-\bar{\bar{\tau}}\bar{q})=p\nabla\cdot\bar{q}-\bar{\bar{\tau}}\nabla\cdot\bar{q} \tag{4.155}$$

The divergence term in the first member represents the energy carried per unit of time and volume, but can be expressed in terms of the work done on the boundary by pressure and viscous stress forces:

$$\iiint_V\nabla\cdot(p\bar{q}-\bar{\bar{\tau}}\bar{q})dV=\oint\!\!\oint_S(p\bar{q}-\bar{\bar{\tau}}\bar{q})\cdot\hat{n}\,dS \tag{4.156}$$

Thus the first member of eqn (4.153) represents the variation of kinetic energy. It results that the second member must correspond to the heat received in excess to an increase of internal energy (first principle of thermodynamics) and:

$$\iiint_V(p\nabla\cdot\bar{q}-\bar{\bar{\tau}}\nabla\cdot\bar{q})dV=\iiint_V\rho\left(\frac{dQ}{dt}-\frac{d\varepsilon}{dt}\right)dV \tag{4.157}$$

1. the first term represents the work done against pressure in case of a dilatation ($\nabla\cdot\bar{q}>0$) or a compression ($\nabla\cdot\bar{q}<0$)—a reversible adiabatic conversion of internal energy into mechanical energy;
2. the second term is the energy converted into heat by viscous friction.

If we introduce the conduction heat flow into V:

$$dQ=-\frac{1}{\rho}\nabla\cdot(k\nabla T)dt \tag{4.158}$$

or

$$\rho\frac{dQ}{dt}=-\nabla\cdot(k\nabla T)=-\nabla\cdot\bar{F} \tag{4.159}$$

The heat generated by radioactivity can be included in $\bar{F}$ and, if we add the electrical power $\bar{E}\cdot\bar{j}$ the equation for the mechanical energy becomes

$$\rho\frac{D}{Dt}\left(\varepsilon+\frac{q^2}{2}\right)+\nabla\cdot(p\bar{q}-\bar{\tau}\bar{q})+\nabla\cdot\bar{F}-\bar{E}\cdot\bar{j}=0 \tag{4.160}$$

returning to eqn (4.144), i.e.

$$-\bar{E}\cdot\bar{j}=\frac{1}{\mu}\bar{B}\frac{\partial\bar{B}}{\partial t}+\nabla\cdot(\bar{E}\wedge\bar{H}) \tag{4.161}$$

and integrating to the whole volume to be considered:

$$\iiint_V\left[\rho\frac{D}{Dt}\left(\varepsilon+\frac{q^2}{2}+\frac{B^2}{2\mu}\right)+\nabla\cdot\left\{\bar{F}+p\bar{q}-\bar{\tau}\bar{q}+\frac{1}{\mu}(\bar{E}\wedge\bar{B})\right\}\right]dV=0 \tag{4.162}$$

which is the *thermodynamic equation of conservation of energy*:

1. cooling corresponds to a decrease of ε (internal energy);
2. $\iiint_V\nabla\cdot p\bar{q}\,dV=\oiint_S p\bar{q}\cdot\hat{n}dS$ work of pressure forces on the surface (4.163)
3. $\iiint_V\nabla\cdot\bar{\tau}\bar{q}\,dV=\oiint_S\bar{\tau}\bar{q}\cdot\hat{n}dS$ work of viscous stress applied on the surface. (4.164)

If we considered only the "mechanical" energy we should write the conservation equation as

$$\iiint_V\left[\rho\frac{D}{Dt}\left(\frac{q^2}{2}\right)+\bar{q}\cdot\nabla p-\bar{q}\cdot(\bar{j}\wedge\bar{B})-\eta\bar{q}\nabla^2\bar{q}\right]dV=0 \tag{4.165}$$

with (including here compressibility effects)

$$\nabla\cdot p\bar{q}=\bar{q}\cdot\nabla p+p\nabla\cdot\bar{q}=\bar{q}\cdot\nabla p+p\frac{1}{V}\frac{DV}{Dt} \tag{4.166}$$

$$=\bar{q}\cdot\nabla p+\rho p\frac{DV}{Dt}$$

$$\nabla\cdot\bar{\tau}\bar{q}=\bar{q}\cdot(\nabla\cdot\bar{\tau})+(\bar{\tau}\,\nabla)\bar{q} \tag{4.167}$$

Let us subtract eqn (4.165) from (4.160), taking (4.148), (4.166) and (4.167) into account:

$$\rho\frac{D\varepsilon}{Dt}+\rho p\frac{DV}{Dt}=\frac{1}{\sigma}|\bar{j}|^2-\nabla\cdot\bar{F}+(\bar{\tau}\,\nabla)\bar{q} \tag{4.168}$$

but

$$d\varepsilon=TdS-PdV \tag{4.169}$$

$$\frac{1}{T}\nabla\cdot\bar{F}=\nabla\cdot\frac{\bar{F}}{T}-\bar{F}\cdot\nabla T^{-1}=\nabla\cdot\frac{\bar{F}}{T}+\frac{\bar{F}\cdot\nabla T}{T^2} \tag{4.170}$$

and, finally

$$\rho\frac{DS}{Dt}+\nabla\cdot\frac{\bar{F}}{T}=\frac{1}{T}\left\{\frac{|\bar{j}|^2}{\sigma}+(\bar{\bar{\tau}}\nabla)\bar{q}-\frac{\bar{F}\cdot\nabla T}{T^2}\right\} \tag{4.171}$$

which is the *equation of conservation of Entropy.*

$\nabla\cdot\frac{\bar{F}}{T}$ is the flow of entropy inside V;

$\frac{\bar{F}\cdot\nabla T}{T^2}$ is the dissipation into heat (negative term);

$\frac{1}{T}\frac{|\bar{j}|^2}{\sigma}$ is the ohmic dissipation coming from the Poynting equation;

$(\bar{\bar{\tau}}\nabla)\bar{q}$ is the viscous dissipation.

Evidently the three equations are not independent, and only two must be taken into account.
Note that in case of mixtures one has to add the term $(\mu/T)\nabla\cdot\bar{i}$ in the second member of the equation of conservation of entropy:

$$\mu\nabla\cdot\bar{i}=\nabla\cdot\mu\bar{i}-\bar{i}\nabla\mu \tag{4.172}$$

and one can combine with the flow of entropy to have one term

$$\frac{1}{T}(\nabla\cdot(\bar{F}-\mu\bar{i})+\bar{i}\nabla\mu) \tag{4.173}$$

(see chapter 2).

Authors disagree considerably upon the amount of ohmic dissipation Q (eqn 4.149): 10^{11} watts (10^{18} erg s^{-1}) according to Bullard and Gellman (1954), 4×10^{12} watts (4×10^{19} erg s^{-1}) for a strong field according to Braginsky (1964), 10^{10} watts (10^{17} erg s^{-1}) for a weak field according to Pekeris *et al.* (1973).

There was much concern about the possibility of the astronomical luni-solar precession regenerating the magnetic field (Malkus, 1963, 1968).

However, Rochester *et al.* (1975) estimate that the power available from precession is not more than 10^8 watts, that is at least one order of magnitude too small.

They concluded that "there is no need to investigate the much more difficult question of whether or not the pattern of flow it induces is regenerative".

4.12 The geodynamo

General considerations

The "dynamo equation" is eqn (4.24) which can be written:

$$\left(\frac{\partial}{\partial t}+\bar{q}\cdot\nabla\right)\bar{B}=\bar{B}\cdot\nabla\bar{q}+\eta^{*}\nabla^{2}\bar{B} \tag{4.174}$$

as

$$\nabla\Lambda(\bar{q}\Lambda\bar{B})=\bar{B}\cdot\nabla\bar{q}-\bar{q}\cdot\nabla\bar{B} \text{ because } \nabla\cdot\bar{q}=0,\ \nabla\cdot\bar{B}=0$$

Solutions for this equation have been obtained for a growing magnetic field $\bar{B}$. However the Navier equation, completed with the Lorentz force:

$$\frac{\partial\bar{q}}{\partial t}+(\bar{q}\cdot\nabla)\bar{q}+2\bar{\Omega}\Lambda\bar{q}=-\frac{1}{\rho}\nabla p+\nabla\phi+\nu\nabla^{2}\bar{q}+\frac{1}{\mu\rho}(\nabla\Lambda\bar{B})\Lambda\bar{B} \tag{4.175}$$

must be *simultaneously* satisfied. Both equations describe the mutual interaction between the velocity field and the magnetic field.

The combination of both equations is the "*geodynamo problem*". The action of the Lorentz force is here essential for the determination of the magnetic field as it puts limits to the flow $\bar{q}$ which maintains $\bar{B}$ against the dissipation $\eta^{*}\nabla^{2}\bar{B}$.

This problem is so difficult that it has not been solved. What the authors do is to prescribe a pattern of motions ($\bar{q}$), satisfying eqn (4.175) and such that, *given a weak initial magnetic field* (the origin of which is unknown), it increases its intensity until the Lorentz force is sufficiently strong to counteract it as a feedback. They do not allow this field to grow indefinitely but produce just the magnetic field as observed outside the conductor. This is called the *kinematic dynamo model.*

The pioneering work of Bullard and Gellman (1954) was developed in that direction.

A *dynamic dynamo theory* should start with a velocity field satisfying the Navier equation completed with Lorentz, Archimedean and Coriolis forces. This is what Braginski (1964 and later) and what Steenbeck, Krause and Rädler (1966) suggested. As an example of the tentative procedure, Busse (1973) proposed to solve the problem in three steps:

1. Solve the Navier equations without magnetic field.
2. Use the resulting flow to solve a kinematic dynamo and calculate the magnetic field which grows with time.
3. Introduce the Lorentz force to reduce the convection and limit the growing of the magnetic field, in this way reaching a stable equilibrium.

The major obstacle in the construction of models comes from the fact that it is impossible to observe the toroidal field which is a necessary component in eqn (4.175) while the poloidal field is known to be only about 4 gauss at the surface of the core (§ 4.2).

It is thus not surprising to meet two different options in the literature:

1. The toroidal field is sensibly larger than the poloidal field and may reach 50 to 300 gauss. In that case the Lorentz and Coriolis forces are nearly equal and high energy is required to counteract the ohmic dissipation which is proportional to B^2. This may be called the *strong field option* and corresponds to what will be called the $\alpha\omega$ type dynamo (see later). Braginsky (1964), Hide (1966) and others adopt this option.

   With a ratio of about 1000 between the conductivities in the core and in the mantle (see figure 2.3), a 300 gauss toroidal field in the core should produce a field of only 0.3 gauss in the mantle. But the conductivity of the upper mantle is even less than that. An argument in favor of this option is that, to reach an explanation of the inversions of a magnetic field, a toroidal field larger than the poloidal field seems to be necessary.
2. The toroidal field is of the same size as the poloidal field, that is 4 to 5 gauss, so that the Lorentz force remains small and does not upset the geostrophic balance: the motions are in the form of Taylor columns. This may be called the *weak field option, it does not require much energy* and corresponds to what will be called the α^2 type dynamo. Busse (1975) has adopted this option.

Anyway the toroidal field could not be too strong otherwise there would be a large ohmic dissipation (one should remember that it is proportional to the square of the magnetic field) which would be incompatible with heat flow measurements (see §2.13). This is, at present, the only constraint one can introduce in these speculations.

We are thus faced with two basic uncertainties:

1. what is the size of the Lorentz force?
2. which is the source of energy maintaining the velocity field $\bar{q}$? (see for example Gubbins, 1974).

Moreover one has to meet the famous *Cowling "antidynamo theorem"* which states that a self generation by axially symmetrical motions is impossible and that a three dimensional mechanism is needed: this three dimensionality involves severe mathematical difficulties.

One has, therefore, to give an estimation of the lack of reflectional symmetry of the motions which can be characterized through the "*density of helicity*" of the velocity field defined as the average value:

$$D = \langle \bar{q} \cdot (\nabla \Lambda \bar{q}) \rangle = \langle \bar{q} \cdot \bar{\omega} \rangle \tag{4.176}$$

the curl $\bar{\omega} = \nabla \Lambda \bar{q}$ is an antisymmetric tensor twice covariant that we can

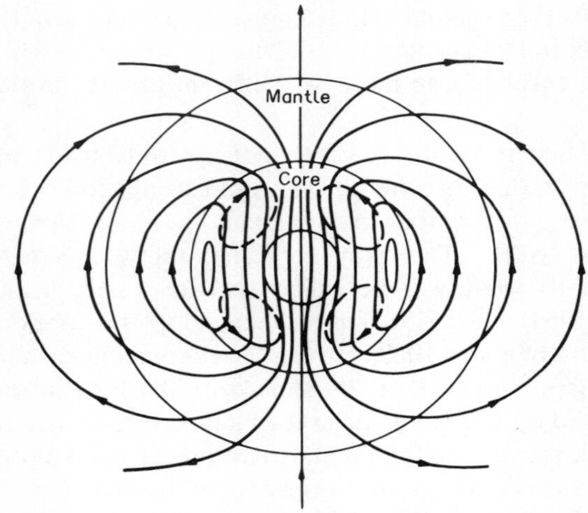

FIGURE 4.9 Poloidal lines of the force of "model Z": the field in the core is nearly parallel to the axis of rotation, with a rather abrupt transition into the potential field in and above the mantle. Dashed curves indicate meridional circulation in the core (after Braginsky). (Stix, 1983).

interpret as a contravariant tensorial density

$$R^k = r_{ij}$$

on the condition that the order of the axis (i, j, k) is prescribed. The product $D = \bar{q} \cdot \bar{\omega}$ is consequently a scalar density, that is a pseudo-tensor, as it changes its sign when we change the axis.

One can give as a simple example the classical circular helice:

$$x = R \cos \Omega t \qquad y = R \sin \Omega t \qquad z = kt$$

for which

$$\dot{x} = -\Omega y \qquad \dot{y} = \Omega x \qquad \dot{z} = k$$

$$r_{11} = 0 \qquad r_{22} = 0 \qquad r_{33} = 2\Omega$$

thus

$$D = \bar{q} \cdot \bar{\omega} = 2k\Omega \tag{4.177}$$

which has the dimensions of an acceleration.

Buoyancy, due to the solidification process at the inner core surface, combined with the Coriolis force obviously generates helicity.

Turbulence in magnetohydrodynamics

The turbulent dynamo mechanism was initially developed by Steenbeck and Krause (1966). We indicate here only the main lines of these rather difficult developments.

The concept of turbulence in magnetohydrodynamics is similar to the concept of turbulence in pure hydrodynamics: one separates the average motion and the random turbulent asymmetric motion for the magnetic field as follows:

$$\bar{q} = \langle \bar{q}_0 \rangle + \bar{q}' \quad \langle \bar{q}' \rangle = 0$$
$$\bar{B} = \langle \bar{B}_0 \rangle + \bar{b}' \quad \langle b' \rangle = 0 \tag{4.178}$$

so that the induction equation gives rise to two equations

$$\frac{\partial \bar{B}_0}{\partial t} = \nabla\Lambda(\bar{q}_0 \Lambda \bar{B}_0) + \nabla\Lambda(\underline{\langle \bar{q}' \Lambda \bar{b}' \rangle}) + \eta^* \nabla^2 \bar{B}_0 \tag{4.179}$$

$$\frac{\partial \bar{b}'}{\partial t} = \nabla\Lambda(\bar{q}_0 \Lambda \bar{b}') + \underline{\nabla\Lambda(\bar{q}' \Lambda \bar{B}_0)} + \nabla\Lambda(\bar{q}' \Lambda \bar{b}') - \nabla\Lambda(\langle \bar{q}' \Lambda \bar{b}' \rangle) + \eta^* \nabla^2 \bar{b}' \tag{4.180}$$

where a new *mean electromotive force* appears, which is produced by the random fluctuations:

$$\bar{E} = \langle \bar{q}' \Lambda \bar{b}' \rangle \tag{4.181}$$

which are of fundamental importance for the dynamo mechanism: this term is clearly analogous to the quadratic term $\langle u'_i u'_j \rangle$ which is the source of the Reynolds tensor in the Navier equations.

In eqn (4.180) the term $\nabla\Lambda(\bar{q}' \Lambda \bar{B}_0)$ acts as a source which generates the fluctuating field $\bar{b}'$. If one supposes that, initially, one had $\bar{b}' = 0$, then the linearity of (4.180) allows one to relate the $\bar{B}_0$ and $\bar{b}'$ fields with a linear relation.

It results that $\bar{E} = \langle \bar{q}' \Lambda \bar{b}' \rangle$ and $\bar{B}_0$ are also linearly related.

As the scale of $\bar{B}_0$ is large it can be expected that a development of $\bar{E}$ in series will be rapidly convergent so that one can tentatively write:

$$\bar{E}_i = \alpha_{ij} \bar{B}_{0j} + \beta_{ijk} \frac{\partial \bar{B}_{0j}}{\partial x_k} + \gamma_{ijkl} \frac{\partial^2 \bar{B}_{0j}}{\partial x_k \partial x_l} + \dots \tag{4.182}$$

$\alpha_{ij}, \beta_{ijk} \dots$ are pseudo-tensors (as $\bar{B}$ is an axial vector and $\bar{E}$ a polar vector), α_{ij} being more exactly a pseudo-scalar as it changes sign for a right-hand (left-hand) transformation.

These pseudo tensors are totally determined by

1. the mean field $\bar{q}_0$
2. the statistical properties of the random turbulent field
3. the parameter η^*

If one subdivides α_{ij} in its symmetric and antisymmetric parts:

$$\alpha_{ij} = \alpha_{ij}^{(s)} - \varepsilon_{ijk} a_k \tag{4.183}$$

$$a_k = -\tfrac{1}{2} \varepsilon_{ijk} \alpha_{ij} \tag{4.184}$$

one observes that the antisymmetric part adds a polar vector $\bar{a}$ to the mean velocity $\bar{q}_0$ because

$$\bar{E}_i = \alpha_{ij}^{(s)} \bar{B}_{0j} + (\bar{a} \Lambda \bar{B}_0)_i + \ldots \tag{4.185}$$

$$\frac{\partial \bar{B}_0}{\partial t} = \nabla \Lambda (\bar{q}_0 \Lambda \bar{B}_0) + \nabla \Lambda (\bar{a} \Lambda \bar{B}_0)_i + \ldots + \eta^* \nabla^2 \bar{B}_0 \tag{4.186}$$

$\bar{q}_0 + a$ is the effective mean velocity.

α appears as a statistical property of the velocity field which is different from zero only if the velocity field $\bar{q}$ has no symmetry by reflexion (Moffat, 1970, p. 152).

Role of the β_{ijk} coefficient

If one considers the term

$$E_i^{(1)} = \beta_{ijk} \frac{\partial B_j}{\partial x_k} \quad (-\beta \nabla \Lambda \bar{B}_0 = \bar{E}^{(1)}) \tag{4.187}$$

it results that

$$\nabla \Lambda \bar{E}^{(1)} = \beta \nabla^2 \bar{B}_0 \quad (\nabla \cdot \bar{B}_0 = 0) \tag{4.188}$$

which shows that β modifies the normal magnetic diffusion by the addition of a turbulent magnetic diffusivity.

One can therefore put

$$\eta^{**} = \eta^* + \beta \tag{4.189}$$

so that the equations can be written

$$\bar{E} = \alpha \bar{B} - \beta (\nabla \Lambda \bar{B}) \tag{4.190}$$

$$\frac{\partial \bar{B}}{\partial t} = \nabla \Lambda (\bar{q} \Lambda \bar{B}) + \alpha (\nabla \Lambda \bar{B}) + \eta^{**} \nabla^2 \bar{B} \tag{4.191}$$

We can thus conclude that:

α introduces a regenerative electromagnetic force

β introduces a dissipative force

The α effect

The so-called α effect appears as a consequence of the small scale turbulent eddies, having helicoidal properties with a preferred sense. This turbulence stimulates an electromotive force which is parallel to the mean magnetic field B and is a function of B and of its derivatives:

$$\bar{\varepsilon} = \alpha \bar{B} - \beta \operatorname{curl} \bar{B} \tag{4.192}$$

a force which has to be included in the Ohm's law expression:

$$\bar{j} = \sigma(\bar{E} + \bar{q}\Lambda\bar{B} + \bar{\varepsilon}) \tag{4.193}$$

α, which is a statistical property of the velocity field, is a pseudo scalar, being negative or positive according to the sense of the helicoidal rotation. It results that $\bar{\varepsilon}$ will be parallel or antiparallel to $\bar{B}$.

The equation of induction then becomes

$$\frac{\partial \bar{B}}{\partial t} - \eta^{**}(\nabla\Lambda(\nabla\Lambda B)) = \nabla\Lambda(\bar{q}\Lambda\bar{B}) + \nabla\Lambda(\alpha\bar{B}) \tag{4.194}$$

where the force $\bar{\varepsilon} = \alpha\bar{B}$ is a source which regenerates and amplifies the magnetic field against the turbulent magnetic viscous dissipation (η^{**}).

The Coriolis force causes anisotropy in the turbulence which necessitates the replacement of α by a tensor α_{ik}.

The dynamo process could therefore appear as the result of an interaction between the poloidal magnetic field and a toroidal field and vice versa. As $\bar{B}$ is solenoidal we can always put

$$\bar{B} = \text{curl curl}\, \bar{r}\psi + \text{curl}\, \bar{r}\phi \tag{4.195}$$

Thus, if a toroidal field exists, the α effect generates a poloidal field which, in turn, generates a toroidal field: it is called *the α^2 dynamo model* where the two fields have about the same order of magnitude. This is *the weak field model.*

Another generation mechanism can be proposed: a differential rotation originating in the Archimedean buoyancy forces. The transport of fluid by thermal convection has necessarily a radial component. As the kinetic momentum in the external layers is greater than in the internal layers, such a radial movement generates a differential rotation, accelerating the internal layers and decelerating the outer layers. This causes a stretching of the frozen magnetic field lines which produce a toroidal field from a poloidal field but not the reverse: this is *the $\alpha\omega$ dynamo model* where the toroidal field is much more important because the stretching of the field lines is a very efficient process. This is *the strong field model.*

This very short summary of the dynamo process clearly shows that the major obstacle in the elaboration of a theory is the lack of information about the movements which take place inside the core.

One obviously needs new kinds of experimental information to constrain the models.

One remark can be made about such constraints: the heating by joule effect is proportional to B^2, it cannot produce too strong a heat flow inside the mantle. This already puts a limit to the toroidal field intensity (a weak field gives $Q \sim 10^9$ to 10^{10} joule s^{-1} while a strong field can give 10^{12} joule s^{-1}).

The joule dissipation is small in the α^2 process but turbulent viscosity is more efficient in that case than in the $\alpha\omega$ mechanism (a strong magnetic field can prevent turbulence).

Busse has developed the α^2 dynamo process in a number of remarkable papers, while Braginsky has essentially worked out the $\alpha\omega$ dynamo process in a number of equally remarkable papers.

For a complete development of these theories we cannot do better than refer the reader to the treatise of H. K. Moffat (1978).

4.13 Fluid movements in the core

Historical background

Elsasser (1946) has considered the case of magnetostatics, i.e. $|\bar{q}| = 0$. Then, equations (4.10) and (4.20) give

$$\operatorname{curl} \bar{B} = \mu\sigma\bar{E} \tag{4.196}$$

or, with (4.5) and (4.17):

$$\nabla^2 A = \operatorname{grad} \operatorname{div} \bar{A} - \operatorname{curl} \operatorname{curl} \bar{A} = -\operatorname{curl} \bar{B} = -\mu\sigma\bar{E} = \mu\sigma\frac{\partial \bar{A}}{\partial t} \tag{4.197}$$

Thus

$$\left.\begin{array}{lr} \text{inside the core} & \nabla^2\bar{A} - \mu\sigma\dfrac{\partial \bar{A}}{\partial t} = 0 \\ \text{outside the core} & \nabla^2\bar{A} = 0 \\ (\text{if } \sigma = 0) & \end{array}\right. \tag{4.198}$$

and, putting

$$\begin{gathered} \bar{A} = \bar{A}_0 e^{-\lambda t} \\ \lambda\mu\sigma = k^2 \end{gathered} \tag{4.199}$$

he obtains the vectorial wave equation of Helmholtz

$$\nabla^2\bar{A}_0 + k^2\bar{A}_0 = 0 \tag{4.200}$$

The solutions of which should yet satisfy the Maxwell equations.

Bullard (1949) considers a rotation

$$\bar{q} = \bar{\omega}\Lambda\bar{r} \tag{4.201}$$

and, from (4.21):

$$\left.\begin{aligned} \operatorname{curl} \bar{B} &= \mu\sigma(\bar{E} + \bar{q}\Lambda\bar{B}) \\ \operatorname{curl} \bar{E} &= -\frac{\partial \bar{B}}{\partial t} \end{aligned}\right\} \tag{4.202}$$

write

$$-\text{curl curl}\,\bar{B} = \mu\sigma\left(\frac{\partial \bar{B}}{\partial t} + \text{curl}\,(\bar{q}\Lambda\bar{B})\right) \tag{4.203}$$

with

$$-\text{curl}\,(\bar{q}\Lambda\bar{B}) = \bar{B}\cdot\text{grad}\,\bar{q} - \bar{q}\,\text{grad}\,\bar{B} \tag{4.204}$$

as $\qquad \text{div}\,\bar{q} = 0 \qquad \text{div}\,\bar{B} = 0$

Let us write

$$\bar{\omega}:\ \begin{pmatrix} \omega\cos\theta \\ -\omega\sin\theta \\ 0 \end{pmatrix} \begin{matrix} \hat{r} \\ \hat{\theta} \\ \hat{\lambda} \end{matrix} \qquad \bar{r}:\ \begin{pmatrix} r \\ 0 \\ 0 \end{pmatrix} \begin{matrix} \hat{r} \\ \hat{\theta} \\ \hat{\lambda} \end{matrix}$$

Then

$$\bar{q} = \bar{\omega}\Lambda\bar{r} = (-\omega\sin\theta r)(\hat{\theta}\Lambda\hat{r}) = +(\omega r\sin\theta)\hat{\lambda} \tag{4.205}$$

$$-\text{curl}(\bar{q}\Lambda\bar{B}) = -\frac{\partial}{r\sin\theta\partial\lambda}(-\omega r\sin\theta\,\bar{B}) = \omega\frac{\partial\bar{B}}{\partial\lambda} \tag{4.206}$$

because $\bar{q}$ has a component in λ only. Finally

$$-\text{curl curl}\,\bar{B} = \mu\sigma\left(\omega\frac{\partial\bar{B}}{\partial\lambda} + \frac{\partial\bar{B}}{\partial t}\right) \tag{4.207}$$

if one looks for solutions in the form

$$\bar{B} = B_0\,e^{im\lambda}\,e^{-\sigma t} \tag{4.208}$$

one has

$$-\text{curl curl}\,\bar{B} = \mu\sigma(i\omega m - \sigma)\bar{B} \tag{4.209}$$

and one should put here

$$\mu\sigma(\sigma - i\omega m) = k^2 \tag{4.210}$$

to write the same vectorial wave equation as (4.200):

$$\nabla^2\bar{B} + k^2\bar{B} = 0 \tag{4.211}$$

Let us take a toroidal solution

$$\bar{T} = \text{curl}(\psi\bar{r}) = -\bar{r}\Lambda\,\text{gr}\bar{\text{a}}\text{d}\,\psi \tag{4.212}$$

and put

$$\text{curl}\,\bar{T} = k\bar{S} \tag{4.213}$$

One sees that

$$\nabla^2\bar{T} + k^2\bar{T} = -\text{curl curl}\,\bar{T} + k^2\bar{T} = -k\,\text{curl}\,\bar{S} + k^2\bar{T} = 0 \tag{4.214}$$

thus

$$\text{curl}\,\bar{S} = k\bar{T} \tag{4.215}$$

Toroidal solutions are such that the curl of a solution is again a solution so that we can couple them in the following way:

$$\overline{B} = \overline{T} + \overline{S} = \text{curl}(\psi(\bar{r}, t)\bar{r}) + \text{curl curl}(\chi(\bar{r}, t)\bar{r}) \tag{4.216}$$

Moreover

$$\begin{aligned} k\overline{S} &= \text{curl curl}(\psi\bar{r}) = -\nabla^2(\psi\bar{r}) + \text{grad div}\,\psi\bar{r} \\ &= k^2\psi\bar{r} + \text{grad div}\,\psi\bar{r} \end{aligned} \tag{4.217}$$

Thus

$$\overline{S} = k\psi\bar{r} + k^{-1}\,\text{grad}\left(\frac{\partial}{\partial r}\psi\bar{r}\right) \tag{4.218}$$

Now let us develop the components of $\overline{T}$ and $\overline{S}$. $\overline{T}$ has no radial component:

$$\bar{r}: \begin{pmatrix} r \\ 0 \\ 0 \end{pmatrix} \qquad \text{grad}\,\psi: \begin{pmatrix} \partial\psi/\partial r \\ \partial\psi/r\partial\theta \\ \partial\psi/r\sin\theta\partial\lambda \end{pmatrix} \tag{4.219}$$

give indeed

$$T_r = 0 \qquad T_\theta = \frac{\partial\psi}{\sin\theta\partial\lambda} \qquad T_\lambda = -\frac{\partial\psi}{\partial\theta} \tag{4.220}$$

But $\overline{S}$ has a radial component, it is a spheroidal mode which can be written, in spherical coordinates:

$$\left.\begin{aligned} S_r &= k\psi r + k^{-1}\frac{\partial^2}{\partial r^2}(\psi r) = k\psi r + n(n+1)(kr)^{-1}\psi \\ S_\theta &= (kr)^{-1}\partial^2(\psi r)/\partial r\partial\theta \\ S_\lambda &= (kr)^{-1}\partial^2(\psi r)/\sin\theta\partial r\partial\lambda \end{aligned}\right\} \tag{4.221}$$

Thus, as

$$\mu\bar{j} = \text{curl}\,\overline{B} = \mu\sigma(\overline{E} + \bar{q}\Lambda\overline{B}) \tag{4.222}$$

it is clear that:

if $\overline{B}$ is toroidal $\quad \bar{j}$ and $\overline{E}$ are spheroidal

if $\overline{B}$ is spheroidal $\quad j$ and $\overline{E}$ are toroidal

A toroidal field is associated with poloidal currents which cannot penetrate an insulating mantle. Hence the toroidal magnetic field must vanish at the core–mantle boundary if the mantle is insulating.

A toroidal velocity field $\bar{q}$ induces a poloidal magnetic field and vice-versa. To satisfy the wave equation

$$\nabla^2\psi + k^2\psi = 0 \tag{4.223}$$

in a sphere, Bullard has chosen Bessel functions of r and surface harmonic functions in θ, λ, defining the stream function

$$\psi_{sn}^m = (kr)^{-1/2} J_{n+1/2}(k_{sn}r) P_n^m(\theta) e^{im\lambda} e^{-\sigma t} \tag{4.224}$$

the k_{sn} being determined by the condition

$$J_{n-1/2}(k_{sn}b) = 0 \tag{4.225}$$

b is the core radius.

It is at present known that kinematic dynamos can exist for a wide variety of fluid motion patterns. Besides the energy budget, the experimental description of such patterns could offer another constraint—to identify the source of mechanical energy.

The difficulty results from the fact that our observations are made quite far from the core boundary, and that the downward extrapolation of the observed field considerably enhances the experimental errors affecting the higher-order terms of the harmonic development (3.26).

Tentative determination of the fluid movements at the top of the core: the "null flux curves"

Kahle *et al.* (1967), Backus (1968), Booker (1969) and more recently Benton (1979, 1981), Benton *et al.* (1979) have tried to obtain at least partial information about the fluid movements at the top of the core by comparing downward extrapolation of the field $\bar{B}$ and its secular variation $\partial \bar{B}/\partial t$ at different epochs.

Their basic idea is that, at time scales of some tens of years (anyway less than a century), the magnetic diffusivity can be neglected so that the induction equation reduces to

$$\frac{\partial \bar{B}}{\partial t} = \text{curl}\,(\bar{q} \wedge \bar{B}) \tag{4.226}$$

so that its radial component at the inner edge of the core mantle boundary layer is*

$$\frac{\partial B_r}{\partial t} + \left[u_\theta \frac{\partial}{r\partial\theta} + u_\lambda \frac{\partial}{r \sin\theta \partial\lambda} \right] B_r = B_r \frac{\partial u}{\partial r} \tag{4.227}$$

because

$$\frac{\partial u_r}{\partial\theta} = \frac{\partial u_r}{\partial\lambda} = 0 \qquad u_r = \bar{q}\cdot\hat{r} = 0 \tag{4.228}$$

at this level, u_θ, u_λ is the horizontal velocity with which the core fluid slips past

* $\dfrac{\partial B_r}{\partial t} + \nabla_S \cdot (B_r \bar{q}_S) = 0$

$\nabla_S \cdot (B_r \bar{q}_S) = \bar{q}_S \cdot \nabla_S B_r - \dfrac{\partial u}{\partial r} B_r$

the mantle (we dropped laminar viscosity in eqn (4.226) which allows such slipping).

It is possible, in this way, to find contours where B_r is zero at the core surface. These are called *null flux curves* (where $\overline{B} \cdot \hat{r} = 0$) (christened by Backus, 1968). The magnetic equator is, of course, a null flux curve but the authors have discovered others, in loop forms; essentially one on the southern extremity of South America, one south-east of South Africa and one east of Japan. Inside each null flux curve, B_r reaches an extremum at which $\partial B_r/\partial\theta$ and $\partial B_r/\partial\lambda$ vanish simultaneously.

A tracking of their motion can be made by computing at different epochs, 5- or 10-year intervals: the null flux curves show a drift but this can be interpreted only in form of *orthogonal components* to the null curve because we have no way to check whether or not there is slipping of the particles along the constant B_r curves: the solution for $\bar{q}$ is therefore not unique.

The loops in the southern hemisphere are drifting to the west, faster than the loop of the northern hemisphere which should imply a "non-solid" rotation. The South Africa loop seems to expand, which may be as a result of upflow, while the South American loop seems to contract, which could correspond to a downflow.

This interpretation rests upon the hypothesis of frozen magnetic lines of force inside the fluid so that the null flux curves appear as *tracers* of a *limited* aspect of the core motion field. The null values $B_r = 0$ are indeed retained by the fluid parcels moving on the core surface so that the null flux lines are real material lines.

A serious limitation to the validity of such a procedure is related to the truncation of the development into Legendre polynomials (4.27). Benton *et al.* (1979) have given several examples with truncations to $n = 8, 9, 10, 11, 12$ and observed that, while the minor details are strongly altered with increasing n, the main features remain relatively stable. Booker found that most of the horizontal variation in B is accounted for by orders $n \leq 6$, which corresponds to a 600 km scale length.

It is also to be remembered that the downward extrapolation seriously amplifies the uncertainties at the higher orders so that the fine details of the magnetic field at the surface of the core cannot be obtained as long as the experimental errors of earth surface measurements are not drastically reduced. This is why up to five null flux contours have been found, which probably reveals the regions where magnetic flux passes out of the core (upwelling of fluid near the surface of the core with concomitant ejection of magnetic flux by diffusion) into the mantle or vice-versa.

4.14 Electric radius of the core

In 1978 R. Hide proposed an elegant method to derive the radius of the core mantle interface from the features of the geomagnetic field which are observed at the Earth's surface.

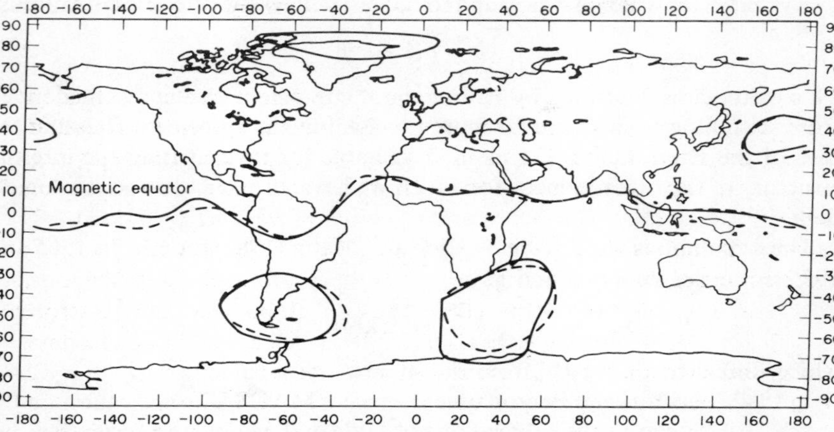

FIGURE 4.10 Null-flux contours ($Z = 0$) on the surface of the core for 1975 (Barraclough *et al.*). $N = 7$ (dashed); $N = 8$ (solid). (from Benton, Muth and Stix, 1979)

We have seen in §4.2 that the number of lines of force per unit normal cross-section area is proportional to the field intensity. If the body is a perfect conductor the total number of lines through any closed contour is therefore constant in time (Parker, 1979) and every line entering the surface must leave it.

If we consider time scales which remain short in comparison with time scales on which ohmic decay can be felt, that is decades, we can treat the core as a perfect conductor. In addition one considers the whole planet as isolated, and its mantle as non-conducting as well as the atmosphere.

The magnetic field $\bar{B}$ is considered to be entirely due to electric currents of density $\bar{j}$ flowing inside the conducting core.

Let us define, with Hide (1978) the total number of intersections of magnetic lines of force with a closed surface $\Sigma \equiv (r = R)$ containing the origin, at time t:

$$N(t; r = R) = \oiint_{\Sigma} |\bar{B} \cdot d\bar{S}| = \int_0^{2\pi} \int_0^{\pi} |B_r(r, \theta, \lambda, t)| r^2 \sin\theta \, d\theta \, d\lambda \quad (4.229)$$

N is expressed in tesla m^2 or weber. It is constant in time at the boundary of the core if the core is a perfect conductor, thus:

$$\frac{\partial N}{\partial t} = 0 \quad \text{at } r = \text{core radius } b \quad (4.230)$$

Hide proposed thus to extrapolate B downwards, from the Earth's surface, for two different epochs of time and to calculate at successive levels inside the mantle (where $\bar{j} = 0$), defined by ε steps:

$$\Delta N(t_1, t_2, r = a(1-\varepsilon)) = N(t_2, r = a(1-\varepsilon)) - N(t_1, r = a(1-\varepsilon)) \quad (4.231)$$

The value ε^* for which ΔN is found to be zero defines the "electric core radius"

$$b = a(1 - \varepsilon^*) \tag{4.232}$$

This method, as described by Hide, cannot compete with the results derived from seismology, which are far more precise, but the agreement found in the case of the Earth makes the method valuable for investigating the internal structure of the other planets, for which we have, of course, no seismological data.

The calculations were made by Hide and Malin (1981) for epochs 1965 and 1975 up to degree $n = 8$ and gave

$$b = 3450 \text{ km}$$

which differs by only 10% from the seismological value.

In 1982, Voorhies and Benton used the recent MAGSAT observations at the epoch 1980 to repeat the calculations of Hide and Malin, and compared this new field to those at different epochs. They got the following results:

1930–1980		3513 km
1940–1980		3430
1950–1980		3445
1965–1980	$n = 8$	3547
	$n = 10$	3433

giving a mean value

$$b = 3484 \pm 48 \text{ km}$$

which does not differ from the 3485 ± 3 km seismological value! Such results support the customary approximation made in geophysics that the mantle is a perfect insulator while the core is a perfect conductor for decade time scales.

Benton (1979) suggests that this property should be used as a constraint for geomagnetic models to produce $\Delta N = 0$ at $r = b$.

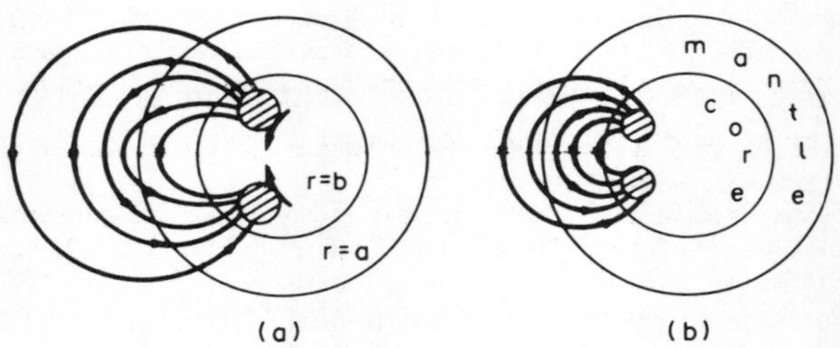

FIGURE 4.11 Schematic diagram from Roberts and Scott (1965) showing the disposition of a particular flux tube leaving and re-entering the core when the two areas of intersection with the core surface are moved together by fluid motions.

Bibliography

Acheson, D. J. (1972) On the hydromagnetic stability of a rotating fluid annulus. *J. Fluid Mech.*, **52**, 529–41.

Acheson, D. J. (1978) Magnetohydrodynamic waves and instabilities in rotating fluids. *Rotating Fluids in Geophysics*. Academic Press, pp. 315–46.

Acheson, D. J. and Hide, R. (1973) Hydromagnetics of rotating fluids. *Rep. Prog. Phys.* **36**, 159–221.

Alfven, H. (1942) Existence of electromagnetic-hydrodynamic waves. *Nature*, **3805**, 405–6.

Allan, D. W. and Bullard, E. C. (1966) The secular variation of the Earth's magnetic field. *Proc. Camb. Phil. Soc.*, **62**, 783–808.

Alldredge, L. R. (1985) More on the alleged 1970 Geomagnetic Jerk. *Phys. Earth Plan. Int.* **39**, 255–63.

Alldredge, L. R. and Braginsky, S. I. (1971) Origin of the geomagnetic field and its secular change. *Trans. XV Gen. Ass. IAGA Bull.* 31, pp. 41–53.

Backus, G. (1958) A class of self-sustaining dissipative spherical dynamos. *Annals of Physics*, **4**, 372–447.

Backus, G. E. (1968) Kinematics of geomagnetic secular variation in a perfectly conducting core. *Phil. Trans. R. Soc. London*, **A263**, 239–66.

Backus, G. E., Parker, R. L. and Zumberge, M. A. (1984) Does the Geoid Drift West? Geopotential Research Mission (GRM). Nasa Conference Publication 2390, 107.

Ball, R. H., Kakla, A. B. and Vestine, E. H. (1969) Determination of surface motions of the Earth's core. *J. Geophys. Res.*, **74**, 3659.

Banks, R. J. (1969) Geomagnetic variations and the electrical conductivity of the upper mantle. *Geophys. J. R. Astr. Soc.* **17**, 457–87.

Batchelor, G. K. (1950) On the spontaneous magnetic field in a conducting liquid in turbulent motion. *Proc. Roy. Soc.*, **201**, 405–16.

Benton, E. R. (1979a) Magnetic probing of planetary interiors. *Phys. Earth Plan. Int.*, **20**, 111–18.

Benton, E. R. (1979b) On fluid circulation around null-flux curves at Earth's core-mantle boundary. *Geophys. Astr. Fluid Dyn.*, **11**, 323–7.

Benton, E. R. (1981a) Inviscid, frozen-flux velocity components at the top of Earth's core from magnetic observations at Earth's surface. Part 1. A new methodology. *Geophys. Astr. Fluid Dyn.*, **18**, 157–74.

Benton, E. R. (1981b) A simple method for determining the vertical growth rate of vertical motion at the top of Earth's outer core. *Phys. Earth Plan. Int.*, **24**, 242–4.

Benton, E. R. (1983) A test for the presence of significant vertical motion in the upper-most layer of Earth's core. IAGA Programme and Abstracts, IUGG General Assembly, Hamburg.

Benton, E. R. (1984) Expected contribution of the Geopotential Research Mission (GRM) to studies of liquid core fluid dynamics. Geopotential Research Mission (GRM). Nasa Conference Publication 2390, 103–4.

Benton, E. R. and Loper, D. E. (1969) On the spin-up of an electrically conducting fluid. I. The unsteady hydromagnetic Ekman–Hartmann Boundary-layer problem *J. Fluid Mech.*, **39**, 561.

Benton, E. R. and Loper, D. E. (1970) On the spin-up of an electrically conducting fluid. II. Hydromagnetic spin-up between infinite flat insulating plates. *J. Fluid Mech.*, **43**, 785.

Benton, E. R. and Muth, L. A. (1979) On the strength of electric currents and zonal magnetic fields at the top of the Earth's core. Methodology and preliminary estimates. *Phys. Earth Plan. Int.*, **20**, 127–33.

Benton, E. R., Muth, L. A. and Stix, M. (1979) Magnetic contour maps at the core–mantle boundary. *J. Geomag. Geoelectr.*, **31**, 615–26.

Bloxham, J. and Gubbins, D. (1984) The magnetic field at the core–mantle boundary. Geopotential Research Mission (GRM). Nasa Conference Publication 2390, 82–4.

Booker, J. R. (1969) Geomagnetic data and core motions. *Proc. Roy. Soc.*, **A309**, 27–40.

Braginskiy, S. I. (1964) Magnetohydrodynamics of the Earth's core. *Geomagnetism Aeronomy*, **4**, 698–712.

Braginsky, S. I. (1976) On the nearly axially-symmetrical model of the hydromagnetic dynamo of the Earth. *Phys. Earth Plan. Int.*, **11**, 191–9.

Braginsky, S. I. (1983) On the theories of the geomagnetic field and its secular variations. *Magnetic Field and the Processes in the Earth's Interior*. Czechoslovak Acad. Sciences, pp. 308–32.

Bullard, E. C. (1949) The magnetic field within the earth *Proc. Roy. Soc., Lond.* **A197**, 433–53.
Bullard, E. C., Freedman, F. R. S. C., Gellman, H. and Nixon, Jo (1950) The westward drift of the Earth's magnetic field. *Phil. Trans.* **A243**, 67–92.
Bullard, E. and Gellman, H. (1954) Homogeneous dynamos and terrestrial magnetism. *Phil. Trans. Roy. Soc., Lond.* **A247** (928), 213–78.
Bullard, E. C. and Gubbins, D. (1977) Generation of magnetic fields by fluid motions of global scale. *Geophys. Astr. Fluid Dyn.* , **8**, 43–56.
Busse, F. H. (1973) Generation of magnetic fields by convection. *J. Fluid Mech.*, **57**(3), 529–44.
Busse, F. H. (1975a) A model of the geodynamo. *Geophys. J. R. Astr. Soc.*, **42**, 437–59.
Busse, F. H. (1975b) Core motions and the geodynamo. *Rev. Geophys. Space Phys.*, **13**(3), 206–41.
Busse, F. H. (1978a) Magnetohydrodynamics of the Earth's dynamo. *Ann. Rev. Fluid Mech.*, **10**, 435–62.
Busse, F. (1978b) Introduction to the theory of geomagnetism. *Rotating Fluids in Geophysics.* Academic Press, pp. 361–87.
Busse, F. H. (1983) Recent developments in the dynamo theory of planetary magnetism. *Ann. Rev. Earth Planet. Sci.* **11**, 241–68.
Busse, F. H. and Miin, S. W. (1979) Spherical dynamos with anisotropic alpha-effect. *Geophys. Astr. Fluid Dyn.*, **14**, 167–81.
Cain, J. C., Schmitz, D. R. and Kluth, C. (1985) Eccentric Geomagnetic Dipole Drift. *Phys. Earth Plan. Int.* **39**, 237–42.
Carrigan, C. R. and Gubbins, D. (1979) The source of the Earth's magnetic field. *Sci. Am.*, **240**, 92–101.
Chandrasekhar, S. (1961) *Hydrodynamic and Hydromagnetic Stability.* Oxford: Clarendon Press.
Collinson, D. W. (1984) Past and present magnetism of the moon. *Geophys. Surv.* **7**, 57–73.
Coulomb, J. (1955) Variation séculaire par convergence òu divergence à la surface du noyau. *Ann. Géophysique*, **T11**, Fasc. 1, pp. 80–2.
Cuong, P. G. and Busse, F. H. (1981) Generation of magnetic fields by convection in a rotating sphere. I. *Phys. Earth Plan. Int.*, **24**, 272–83.
Cupal, I. (1983) Basic Equations and the method in the nearly symmetric dynamo problem. *Magnetic Field and the Processes in the Earth's Interior.* Czechoslovak Acad. Sciences, pp. 332–9.
Currie, R. G. (1967) Magnetic shielding properties of the Earth's mantle *J. Geophys. Res.*, **72**, 2623–33.
Currie, R. G. (1968) Geomagnetic spectrum on internal origin and lower mantle conductivity. *J. Geoph. Res.* **73**, 2779–86.
Elsasser, W. M. (1941) A statistical analysis of the earth's internal magnetic field. *Phys. Review*, **60**, 876–83.
Elsasser, W. M. (1946) Induction effects in terrestrial magnetism. Part II. The secular variation. *Phys. Review*, **70**, 202–12.
Elsasser, W. M. (1946) Induction effects in terrestrial magnetism *Phys. Rev.* **69**(3–4), 106–16.
Elsasser, W. M. (1947) Induction effects in terrestrial magnetism. Part III. Electric modes. *Phys. Rev.* **72**(9), 821–33.
Elsasser, W. M. (1950) Causes of motions in the earth's core *Trans. Am. Geophys. Union*, **31**(3), 454–62.
Elsasser, W. M. (1956) Hydromagnetic dynamo theory. *Rev. Mod. Phys.*, **28**, 135–63.
Fearn, D. R. and Proctor, M. R. E. (1983a) The stabilizing role of differential rotation on hydromagnetic waves. *J. Fluid Mech.* **128**, 21–36.
Fearn, D. R. and Proctor, M. R. E. (1983b) Hydromagnetic waves in a differentially rotating sphere. *J. Fluid Mech.*, **128**, 1–20.
Fearn, D. R. and Proctor, M. R. E. (1984) Self-consistent dynamo models driven by hydromagnetic instabilities. *Phys. Earth Plan. Int.*, **36**, 78–84.
Frederiksen, J. S. and Bell, R. C. (1984) Energy and entropy evolution of interacting internal gravity waves and turbulence. *Geophys. Astr. Fluid Dyn.*, **28**, 171–203.
Garland, G. D. (1981) The significance of terrestrial electrical conductivity variations. *Ann. Rev. Earth Planet. Sci.* **9**, 147–74.
Gilman, P. A. (1971) Instabilities of the Ekman–Hartmann boundary layer. *Phys. Fluids*, **14**(1), 7–12.

Gilman, P. A. and Benton, R. (1968) Influence of an axial magnetic field on the steady linear Ekman boundary layer. *Phys. Fluids*, **11**(11), 2397–401.

Golovkov, V. P. (1983) General analysis of secular variations. *Magnetic Field and the Processes in the Earth's Interior*. Czechoslovak Acad. Sciences, pp. 396–414.

Golovkov, V. P., Kulanin, N. V. and Cherevko, T. N. (1983) Westward drift. *Magnetic Field and the Processes in the Earth's Interior*. Czechoslovak Acad. Sciences, pp. 414–20.

Gubbins, D. (1974) Theories of the geomagnetic and solar dynamos. *Rev. Geophys. Space Phys.*, **12**, 137–54.

Gubbins, D. (1975) Can the Earth's magnetic field be sustained by core oscillations? *Geophys. Res. Lett.*, **2**(9), 409–11.

Gubbins, D. (1976) Observational constraints on the generation process of the Earth's magnetic field. *Geophys. J. R. Astr. Soc.*, **47**, 19–39.

Harrison, C. G. A. (1984) The source of the intermediate wavelength component of the earth's magnetic field. Geopotential Research Mission (GRM). Nasa Conference Publication 2390, 105–6.

Hart, J. E. (1984) Alternative experiments using the geophysical fluid flow cell. Nasa Contractor Report 3766.

Hartmann, J. (1937) Theory of the laminar flow of an electrically conductive liquid in a homogeneous magnetic field. *Kgl. Danske Vid. Selsk. Math. Fys Meddelelser*, **XV**(6), 1–28.

Hartmann, J. and Lazarus, F. (1937) Experimental investigations on the flow of mercury in a homogeneous magnetic field. *Kgl. Danske Vid. Selsk. Math. Fys. Meddelelser*, **XV**(7), 1–45.

Hide, R. (1955) Waves in a heavy, viscous, incompressible, electrically conducting fluid of variable density, in the presence of a magnetic field. *Proc. Roy. Soc. Lond.* **A233**, 376–96.

Hide, R. (1966) Free hydromagnetic oscillations of the Earth's core and the theory of the geomagnetic secular variation. *Phil. Trans. Roy. Soc. Lond.* **A259**, 615–47.

Hide, R. (1967) Motions of the Earth's core and mantle, and variations of the main geomagnetic field. *Science*, **157**, 55–7.

Hide, R. (1969a) On hydromagnetic waves in a stratified rotating incompressible fluid. *J. Fluid Mech.*, **39**(2), 283–7.

Hide, R. (1969b) Dynamics of the atmospheres of the major planets with an appendix on the viscous boundary layer at the rigid bounding surface of an electrically-conducting rotating fluid in the presence of a magnetic field. *J. Atmos. Sci.*, **26**, 841–51.

Hide, R. (1970) On the Earth's core–mantle interface. *Q. J. R. Meteor. Soc.*, **96**(410), 579–87.

Hide, R. (1978) How to locate the electrically-conducting fluid core of a planet from external magnetic observations. *Nature*, **271**, 640–1.

Hide, R. (1979) Dynamo theorems. *Geophys. Astr. Fluid Dyn.*, **14**, 183–6.

Hide, R. (1981a) Self-exciting dynamos and geomagnetic polarity changes. *Nature*, **293**, 728.

Hide, R. (1981b) The magnetic flux linkage of a moving medium: a theorem and geophysical applications. *J. Geophys. Res.*, **86**(B12), 11,681–7.

Hide, R. (1983) The magnetic analogue of Ertel's potential vorticity theorem. *Ann. Geophys.*, **1**, 59–60.

Hide, R. (1984) Rotating fluids in geophysics and planetary physics. I. U. G. G. Chronicle, No. 167–168, pp. 224–236.

Hide, R. and Malin, S. R. C. (1970) Novel correlations between global features of the Earth's gravitational and magnetic fields. *Nature*, **225**, 605–9.

Hide, R. and Malin, S. R. C. (1981) On the determination of the size of the Earth's core from observations of the geomagnetic secular variation. *Proc. R. Soc. Lond.*, **A374**, 15–33.

Hide, R. and Roberts, P. H. (1960) Hydromagnetic flow due to an oscillating plane. *Rev. Mod. Phys.*, **32**, 799–806.

Hide, R. and Roberts, P. H. (1961) The origin of the main geomagnetic field. *Phys. Chem. Earth*, **4**, 27–98.

Hide, R. and Roberts, P. H. (1962) Some elementary problems in magneto-hydrodynamics. *Adv. Appl. Mech.* **7**, 215–316.

Hide, R. and Stewartson, K. (1972) Hydromagnetic oscillations of the Earth's core. *Rev. Geophys. Space Phys.* **10**(2), 579–98.

Ierley, G. R. (1984) Theoretical estimates of the westward drift. *Phys. Earth. Plan. Int.*, **36**, 43–8.

Ingham, D. B. (1969) Magnetohydrodynamic flow in a container. *Phys. Fluids*, **12**(2), 389–96.

Jones, C. A. (1981) Model equations for the solar dynamo. *Solar Magnetism*, 159–70.

Kahle, A. B. (1969) Prediction of geomagnetic secular change confirmed. *Nature*, **223**, 165.
Kahle, A. B., Ball, R. H. and Vestine, E. H. (1967) Comparison of estimates of surface fluid motions of the Earth's core for various epochs. *J. Geophys. Res.*, **72**(19), 4917–25.
Kahle, A. B., Vestine, E. H. and Ball, R. H. (1969) Estimated surface motions of the Earth's core. *J. Geophys. Res.*, **72**(3), 1095–1108.
Kerridge, D. J. and Barraclough, D. R. (1985) Evidence for geomagnetic jerks from 1931 to 1971. *Phys. Earth Plan. Int.* **39**, 228–36.
Knapp, D. G. (1955) The synthesis of external magnetic fields by means of radial internal dipoles. *Ann. Geophys.*, T11, Fasc. 1, pp. 83–5.
Krause, F. and Radler, K. H. (1983) On the theory of the geomagnetic dynamo based on mean field electrodynamics. *Magnetic Field and the Processes in the Earth's Interior*. Czechoslovak Acad. Sciences, pp. 339–61.
Kropachev, E. P., Gorskov, S. N. and Serebrianaya, P. M. (1983) Calculations of some simple models of the steady-state kinematic dynamo. *Magnetic Field and the Processes in the Earth's Interior*. Czechoslovak Acad. Sciences, pp. 362–7.
Kumar, S. and Roberts, P. H. (1975) A three-dimensional kinematic dynamo. *Proc. R. Soc. Lond.*, **A344**, 235–58.
Langel, R. A. and Estes, R. H. (1982) A geomagnetic field spectrum. *Geophys. Res. Lett.*, **9**(4), 250–3.
Langel, R. A. and Estes, R. H. (1985) The near-Earth magnetic field at 1980 determined from MAGSAT data. *J. Geophys. Res.*, **90**, 2495–509.
Lehnert, B. (1954) Magnetohydrodynamic waves under the action of the Coriolis force. *Astrophys. J.*, **119**, 647–54.
Le Mouel, J. L. and Courtillot, V. (1981) Core motions, electromagnetic core–mantle coupling and variations in the Earth's rotation: new constraints from geomagnetic secular variation impulses. *Phys. Earth Plan. Int.*, **24**, 236–41.
Le Mouël, J. L., Gire, C. and Madden, T. (1985) Motions at core surface in the geostrophic approximation. *Phys. Earth Plan. Int.* **39**, 270–87.
Leroy, R. and Hurwitz, L. (1964) Radial dipoles as the sources of the Earth's main magnetic field. *J. Geophys. Res.*, **69**(12), 2631–9.
Lilley, F. E. M. (1970) On kinematic dynamos. *Proc. Roy. Soc. Lond.*, **A316**, 153–67.
Loper, D. E. (1978a) The gravitationally powered dynamo. *Geophys. J. R. Astr. Soc.*, **54**, 389–404.
Loper, D. E. (1978b) Some thermal consequences of a gravitationally powered dynamo. *J. Geophys. Res.*, **83**, 5961–70.
Loper, D. E. and Benton, E. R. (1970) On the spin-up of an electrically conducting fluid. Part 2. Hydromagnetic spin-up between infinite flat insulating plates. *J. Fluid. Mech.*, **43**(4), 785–99.
Lowes, F. J. (1966) Mean-square values on sphere of spherical harmonic vector fields. *J. Geophys. Res.*, **71**, 2179.
Lowes, F. J. (1970) Possible evidence on core evolution from geomagnetic dynamo theories. *Phys. Earth Plan. Int.*, **2**, 382.
Lowes, F. J. (1974) Spatial power spectrum of the main geomagnetic field, and extrapolation to the core. *Geophys. J. R. Astr. Soc.*, **36**, 717–30.
Lowes, F. J. (1984) The geomagnetic dynamo—elementary energetics and thermodynamics. *Geophys. Surv.*, **7**, 91–105.
Malkus, W. V. R. (1963) Precessional torques as the cause of geomagnetism. *J. Geophys. Res.*, **68**, 2871–86.
Malkus, W. V. R. (1967) Hydromagnetic planetary waves. *J. Fluid Mech.*, **28**, 793–802.
Malkus, W. V. R. (1968) Precession of the Earth as the cause of geomagnetism, *Science*, **160**, 259–64.
McDonald, K. L. (1957) Penetration of the geomagnetic secular field through a mantle with variable conductivity. *J. Geoph. Res.* **62**, 117–41.
Moffat, H. K. (1970) Dynamo action associated with random inertial waves in a rotating conducting fluid. *J. Fluid Mech.*, **44**, 705–19.
Moffat, H. K. (1978) *Magnetic field generation in electrically conducting fluids*. Cambridge University Press.
Mullan, D. J. (1973) Earthquake waves and the geomagnetic dynamo. *Science*, **181**, 553.
Muth, L. A. and Benton, E. R. (1981) On the frozen flux velocity at the surface of Earth's core necessary to account for the poloidal main magnetic field and its secular variation. *Phys. Earth Plan. Int.*, **24**, 245–52.

Parker, E. N. (1955) Hydromagnetic dynamo models. *Astrophys. J.*, **122**, 293–314.
Parker, E. N. (1979) *Cosmical Magnetic Fields.* Clarendon Press, Oxford.
Parker, R. L. (1970) The inverse problem of electrical conductivity in the mantle. *Geophys. J. R. Astr. Soc.* **22**, 121–38.
Peddie, N. W. (1981) International geomagnetic reference field 1980. A report by IAGA Division I Working Group I. *J. Geomag. Geoelectr.*, **33**, 607–11.
Pekeris, C. L., Accad, Y. and Shkoller, B. (1973) Kinematic dynamos and the Earth's magnetic field. *Phil. Trans. Roy. Soc. Lond.* **A275**, 425–61.
Reitz, J. R. and Milford, F. J. (1962) *Foundations of electromagnetic theory.* page 234. Addison Wesley Publ. Co.
Roberts, P. H. (1967) Singularities of Hartmann layers. *Proc. Roy. Soc. Lond.*, **300**, 94–107.
Roberts, P. H. and Lowes, F. J. (1961) Earth currents of deep internal origin. *J. Geophys. Res.* **66**, 1243–54.
Roberts, P. H. and Scott, S. (1965) On analysis of the secular variation. *J. Geomagn. Geoelectr.*, **17**(2), 137–51.
Roberts, P. H. and Soward, A. M. (1972) Magnetohydrodynamics of the Earth's core. *Ann. Rev. Fluid Mech.*, **4**, 117–53.
Rochester, M. G., Jacobs, J. A., Smylie, D. E. and Chong, K. F. (1975) Can precession power the geomagnetic dynamo? *Geophys. J. R. Astr. Soc.*, **43**, 661–78.
Shercliff, J. A. (1953) Steady motion of conducting fluids in pipes under transverse magnetic fields. *Proc. Camb. Phil. Soc.*, **49**, 136–44.
Shercliff, J. A. (1956) The flow of conducting fluids in circular pipes under transverse magnetic fields. *J. Fluid. Mech.*, **1**, 644–66.
Shercliff, J. A. (1962) Magnetohydrodynamic pipe flow. Part 2. High Hartmann number. *J. Fluid Mech.*, **13**, 513–18.
Shure, L., Parker, R. L. and Langel, R. A. (1984) Separation of core and crustal magnetic field sources. Geopotential Research Mission (GRM). Nasa Conference Publication 2390, 80–1.
Siran, G. (1974a) The effect of the electric conductivity of the Earth's mantle on the Ekman–Hartman hydromagnetic boundary layer and on the magnetic diffusion region. Part I. The limiting treatment of the problem in connection with the conditions in the Earth's core. *Studia Geophys. Geod.*, No. 3, Series 18, pp. 248–58.
Siran, G. (1974b) The effect of the electric conductivity of the Earth's mantle on the Ekman–Hartman hydromagnetic boundary layer and the magnetic diffusion region. Part II. Approximation of the general solutions with respect to the conditions in the Earth's core. *Studia Geophys. Geod.*, No. 4, Series 18, pp. 359–66.
Siran, G. and Brestensky, J. (1983) The effect of the electric conductivity of the Earth's mantle on the magnetohydrodynamic processes in the Earth's core. *Magnetic Field and the Processes in the Earth's Interior*, Czechoslovak Acad. Sciences, pp. 373–82.
Smylie, D. E. (1965) Magnetic diffusion in a spherically-symmetric conducting mantle. *Geophys. J. R. Astr. Soc.*, **9**, 169–84.
Soward, A. M. (1974) A convection-driven dynamo. I. The weak field case. *Phil. Trans. Roy. Soc. Lond.* **A275**, 611–46.
Soward, A. M. (1978) The kinematic dynamo problem. *Rotating Fluids in Geophysics.* Academic Press, pp. 352–9.
Steenbeck, M. and Krause, F. (1966) Erklarung Stellarer und Planetarer Magnetfelder Durch Einen Turbulenzbedingten Dynamomechanismus. *Z. Naturforsch.*, **21A**, 1285–96.
Steenbeck, M., Krause, F. and Radler, K. H. (1966) Berechnung der Mittleren Lorentz Feldstarke VXB. fur ein Elektrisch Leitendes Medium in Turbulenter, durch Coriolis-krafte Beeinflusster Bewegung. *Z. Naturforsch.*, **21A**, 369–76.
Stewartson, K. (1957) The dispersion of a current on the surface of a highly conducting fluid. *Proc. Camb. Phil. Soc.*, **53**, 774–5.
Stewartson, K. (1967) Slow oscillations of fluid in a rotating cavity in the presence of a toroidal magnetic fluid. *Proc. Roy. Soc. Lond.* **A299**, 173–87.
Stewartson, K. (1978) Waves. *Rotating Fluids in Geophysics.* Academic Press, pp. 67–101.
Szeto, A. M. K. and Cannon, W. H. (1985) On the separation of core and crustal contributions to the geomagnetic field. *Geophys. J. R. Astr. Soc.*, **82**, 319–29.
Taylor, J. B. (1963) The magneto-hydrodynamics of a rotating fluid and the Earth's dynamo problem. *Proc. Roy. Soc. Lond.*, **A274**, 274–83.

Todoeschuck, J. P. and Rochester, M. G. (1980) The effect of compressible flow on anti-dynamo theorems. *Nature*, **284**, 250–1.
Vestine, E. H. (1953) On variations of the geomagnetic field, fluid motions and the rate of the Earth's rotation. *J. Geophys. Res.*, **58**, 127–45.
Vestine, E. H. and Kahle, A. B. (1968) The westward drift and geomagnetic secular change. *Geophys. J. R. Astr. Soc.*, **15**, 29–37.
Vestine, E. H., Ball, R. H. and Kahle, A. B. (1967) Nature of surface flow in the Earth's central core. *J. Geophys. Res.*, **72**(19), 4927–35.
Vestine, E. H., Sibley, W. L., Kern, J. W. and Carlstedt, J. L. (1963) Integral and spherical-harmonic analyses of the geomagnetic field for 1955.0, Part 1. *J. Geomag. Geoelectr.*, **15**, 47–89.
Voorhies, C. V. (1984) Some anticipated contributions to core fluid dynamics from the GRM. Geopotential Research Mission (GRM). Nasa Conference Publication 2390, 85–8.
Voorhies, C. V. and Benton, E. R. (1982) Pole-strength of the Earth from MAGSAT and magnetic determination of the core radius. *Geophys. Res. Lett.*, **9**, 258–61.
Weiss, N. O. (1978) The magnetohydrodynamics of rotating fluids. Basic theory and simple consequences. *Rotating Fluids in Geophysics*. Academic Press, pp. 295–313.
Whaler, K. A. (1980) Does the whole of the Earth's core convect? *Nature*, **287**, 528–30.
Yukutake, T. (1981) A stratified core motion inferred from geomagnetic secular variations. *Phys. Earth Plan. Int.*, **24**, 253–8.

CHAPTER 5

Geophysical Implications in the Earth's Core

5.1 Review summary of the different kinds of waves in the core: classification and terminology

To generate a waveform movement, there must be a *restoring force* giving rise to oscillations when the disturbed particles overshoot their equilibrium position. The fluid mixture which constitutes the Earth's core being stratified, gravitating, rotating and pervaded by a magnetic field the possible restoring forces are, correspondingly, the Archimedean force, the Coriolis force and the Lorentz force. But these forces are weak compared to the elastic force due to the resistance to compression and dilatation which is responsible for the "free oscillations" (§1.24). Therefore, Archimedean, Coriolis and magnetic oscillations will be much slower than the gravest free oscillation (period 54 minutes). This is why these long period oscillations are often called "undertones".

The oscillations resulting from Archimedean buoyancy force are called *internal*; those resulting from the Coriolis force are called *inertial* and those resulting from the Lorentz force are the *magnetic* waves or Alfvén waves. A combination of all three restoring forces would generate *MAC* waves (Magnetic, Archimedean, Coriolis).

The effect of the Coriolis force, which is not strong at short period, increases greatly with the period of the undertones. Similarly the Lorentz force is unimportant at periods less than 1 year. Coriolis force produces helicity of the motions. Stratification and rotation constraints are similar, creating mathematical analogies respectively governed by the Coriolis inertial frequency and by the Brunt–Väisälä buoyancy frequency (Veronis, 1970). The inverse rotational Froude number $A = N^2/4\Omega^2$ is a measure of their relative importance on the oscillatory motions.

The interaction of these two forces in a rotating shell of fluid is not clearly understood and, in this context one can speak of "*inertial gravity waves*".

Inertial waves or planetary waves

This is the case when

$$|\bar{\Omega}| \neq 0, \qquad N = 0, \qquad |\bar{B}| = 0 \tag{5.1}$$

There is no energy other than kinetic: in a more restrictive way, the energy comes from rotation. The Coriolis force here takes the role of a restoring force conferring to the medium a "pseudo-elasticity".

The Coriolis force increases with the latitude and this effect generates at large planetary scale "Rossby waves" which are travelling towards the West. The geostrophic mode, the Rossby–Haurwitz waves, the Laplace oscillations of the second class are all inertial modes.

Their highest angular frequency being twice the rotation frequency, 2Ω, *their periods are in the range of several days.** They are governed by the Poincaré differential equation which becomes hyperbolic because of the rotation (equations (3.97) and (3.108) would be elliptic if $\Omega = 0$). It should be noted also that the Coriolis force do not work and can only change the direction of currents but cannot drive them against dissipation. Aldridge and Toomre succeeded in exciting such inertial waves in a rotating spherical container and in experimentally measuring the resonant modes of oscillations.

Internal waves or gravity waves

This is the case when

$$N \neq 0 \qquad |\bar{\Omega}| = 0 \qquad |\bar{B}| = 0 \tag{5.2}$$

the restoring force is the buoyancy, Archimedean force. This force does mechanical work but the resulting convection restores an equilibrium where $N = 0$ so that energy must be supplied to maintain this convection. Laplace oscillations of the first class are internal waves. Other examples are solitary waves, or pressure jumps.

If we consider a two-dimensional, non-rotating ($f = 2\Omega \sin \phi = 0$) medium, equation (3.164) simplifies as follows:

$$\frac{\partial^2 w}{\partial z^2} - \left(\frac{N^2}{\omega_i^2} - 1\right)\frac{\partial^2 w}{\partial x^2} = 0$$

and, if $w = w(z) \exp i\,(kx - \omega_i t)$, it gives

$$\frac{\partial^2 w}{\partial z^2} + \left(\frac{N^2}{\omega_i^2} - 1\right)k^2 w = 0$$

which shows that:

if $\omega_i < N$ there are wave-like solutions and energy can propagate away from the region where it is generated;

if $\omega_i > N$ the disturbance remains local and falls off exponentially away from the source.

* Longuet–Higgins (1966) showed that for an ocean covering one hemisphere the shortest period is 1.624 day. The periods are longer for basins of lesser extent.

The Brunt–Väisälä frequency thus appears as the upper limit of frequency for which wave motions can exist in a stratified fluid.

If N is variable, waves will be trapped within the layer where it exceeds the imposed frequency.

The problem becomes complicated when the Brunt–Väisälä period is close to the Coriolis period, which seems to be the case especially in the Earth's core. Oscillations are then referred as *inertial gravity waves.*

Theories have recently been developed (Friedlander, 1985) which propose a stably stratified outer layer ($\sim$ 70 km thick, according to Fearn and Loper, 1981) at the core–mantle boundary and a weakly unstable stratification in the deeper parts of the core (but see figure 1.26).

Hydromagnetic waves or Alfvén waves

This is the case when:

$$|\bar{B}| \neq 0 \qquad |\bar{\Omega}| = 0 \qquad N = 0 \tag{5.3}$$

The Lorentz force can be interpreted in terms of Maxwell stress and Maxwell isotropic pressure. Such magnetic forces here act as restoring forces to generate transverse and non dispersive waves travelling along the lines of force with a low speed, equal to:

$$a = H(\mu/\rho)^{1/2} = B(1/\mu\rho)^{1/2}$$

that is the Alfvén speed.

An image of such waves is to consider the tubes of force, frozen into the fluid, as acting like stretched elastic strings. It is easy to see that a speed $a = 4 \times 10^{-3}\,\mathrm{m\,s^{-1}}$ corresponds to $B = 4 \times 10^{-4}$ tesla, $\rho = 1.1 \times 10^{4}\,\mathrm{kg\,m^{-3}}$. Their periods are decades like the periods observed in the geomagnetic field ($\sim$ 60 years).

MAC waves = Magnetic, Archimedean, Coriolis

These waves correspond evidently to the general situation

$$|\bar{B}| \neq 0 \qquad N \neq 0 \qquad |\bar{\Omega}| \neq 0 \tag{5.4}$$

and obey the general equations:

$$\left\{\begin{aligned}
&\frac{\partial \bar{q}}{\partial t} + (\bar{q}\cdot\nabla)\bar{q} + 2\bar{\Omega}\Lambda\bar{q} = \rho^{-1}\nabla p + \nabla\phi + \nu\nabla^2\bar{q} + (\mu\rho)^{-1}(\nabla\Lambda\bar{B})\Lambda\bar{B} \\
&\frac{\partial \bar{B}}{\partial t} = \nabla\Lambda(\bar{q}\Lambda\bar{B}) + \eta^*\nabla^2\bar{B} \\
&\nabla\cdot\bar{q} = 0 \\
&\nabla\cdot\bar{B} = 0
\end{aligned}\right. \tag{5.5}$$

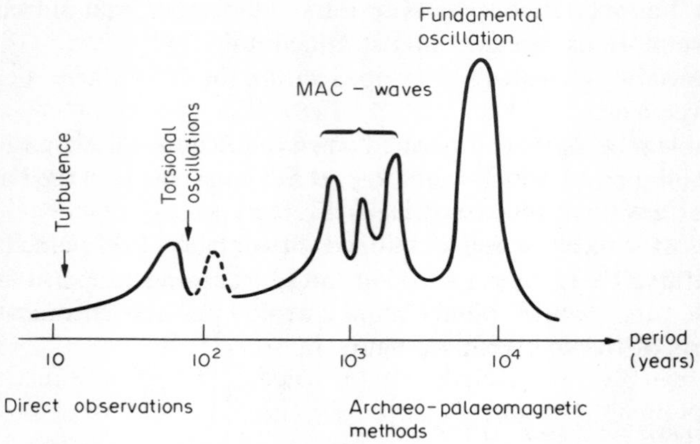

FIGURE 5.1 Schematic picture of the spectrum of geomagnetic field variations according to Braginsky (1971).

and are thus shear waves. In these equations $\bar{q}$, $\overline{B}$, p, ρ are small adiabatic perturbations of equilibrium. N and 2Ω play the roles of limit frequencies in stratified rotating fluids.

Such waves have very long periods ($\sim 10^3$ years), depending upon the strength of the toroidal magnetic field (Hide, 1966).

They can be excited by the Archimedean force and opposed by the stabilizing effect of the magnetic forces acting like elastic forces (Maxwell magnetic stress) but MAC waves are non-diffusive free oscillations, dominated by a balance between Lorentz and Coriolis forces. They have about the right frequency for secular variation, provided the core contains a strong toroidal field.

We could call this a magnetostrophic equilibrium just as the equilibrium between pressure and Coriolis force is called geostrophic. It is also worth while to point out that gravity and magnetic field introduce each a preferential direction.

Remark

Wave motions can probably be excited in the core by strong earthquakes, also by the precession and the nutations, the polar motion, the tides of the boundary (~ 9 cm), even meteorite impacts but chemical and thermal buoyancy seem to be by far the most efficient factors.

5.2 Possible sources of mechanical energy to entertain convection in the Earth's core

The energy is supplied by the work done against the Lorentz force by fluid motions. According to Jacobs (1975, p. 267) 5×10^{16} erg s^{-1} are needed to sustain the convection.

1. Differential precession between core and mantle was suggested by Malkus but ruled out by Rochester *et al.* (1975), but inertial modes could be excited by precession nutations.
2. Thermal convection due to radioactive heating by potassium 40 or uranium 238. Presence of K is advocated by geochemists because of its affinity for S and because the mantle is depleted from K. According to Verhoogen (1973) K forms stable components with S. If S is really the light element in the core, there must be also some K and consequently its isotope K^{40} (lifetime 1.3×10^9 year) must be present. Verhoogen estimates that 2×10^{12} cal s^{-1} ($\sim 10^{13}$ watts) would be available with 0.1 % K in the core. Ringwood contests this possibility.

   Bukowinski and Hauser (1980) calculate that at sufficiently high pressure the heavy alkali metals may undergo a change in electronic configuration which would considerably alter their chemical properties. This may enhance incorporation of radioactive elements into the core during the early evolution of the Earth.

   More recently Feber *et al.* (1984) have proposed uranium 238 as a possible radioactive element on the basis of experimental measurements of the binary steel–UO_2 which, above 3120 K exhibits a two-phase liquid mixture (the temperature at the core interface is 3157 K according to Stacey—cf. Table 2.1). These authors consider that, if at cosmic abundance, there should be 1.45×10^{18} g of U in the core which should provide 2.7×10^{17} ergs s^{-1} sufficient to sustain convection.
3. Thermal convection by release of latent heat of solidification at the inner core surface which is given by the Clausius Clapeyron equation (2.68). This depends from the volume decrease. The efficiency of thermal processes is low because 3×10^{12} watts are simply evacuated by conduction and do not cause convection (see eqn 2.98). Moreover not too much heat can be convected unless the heat flow across the mantle is larger than what is really observed, and then could produce melting in the mantle.
4. Non-thermal convection maintained by the potential gravitational energy released by precipitation of Fe on the inner core while the lighter solution moves upward by buoyancy effect (Verhoogen, 1961). This essentially depends upon the density contrast (see Table 1.4) at the inner core surface and acts as long as the mixture is more metallic than the eutectic: it should stop when the eutectic composition is reached. This explains that planets with a conducting metallic liquid core may have no magnetic field if the mixture is just eutectic, except if the latent heat of crystallization is important (Häge and Müller, 1979). This is at present the hypothesis most in favour.

We do not know if the entire core is subject to convection or if only part of it (top or bottom, or both) is involved in the process (see fig. 1.6 showing the Brunt–Väisälä frequency).

5.3 Gravitational potential energy liberated during the formation of the inner core

If the liquid core is slightly more metallic than the eutectic composition Fe can precipitate to form the "solid" inner core while the mixture will slowly evolve towards the eutectic composition.

If the gravitational potential at a point inside the core is ϕ, the gravitational energy is:

$$E = -\int_V \phi\, dm = -\int_V \left(\frac{G \int_0^r \rho x^2\, d\mu\, d\lambda\, dx}{r}\right) \rho r^2\, d\mu\, d\lambda\, dr \tag{5.6}$$

($\mu = \cos\theta$, θ is the colatitude).

Let us simply consider the case of an homogeneous sphere of radius a:

$$E = -\int_V \left(\frac{4\pi G\rho r^2}{3}\right) \rho r^2\, d\mu\, d\lambda\, dr = -\frac{(4\pi\rho)^2}{3} G \frac{a^5}{5}$$

as

$$M = \frac{4\pi\rho a^3}{3}$$

it results that

$$\boxed{E = -\frac{3}{5}\frac{GM^2}{a}} \tag{5.7}$$

An homogeneous core of mean density 10.59 gives

$$E = -4.020 \times 10^{38} \text{ ergs}$$

while an homogeneous core of mean density 10.50 with an inner core of density 12.5, radius 1220 km gives

$$E = -3{,}987 \times 10^{38} \text{ ergs}$$

Thus, when evolving from the first one to the second one a potential gravitational energy

$$dE = 3.3 \times 10^{36} \text{ ergs} = 3.3 \times 10^{29} \text{ joules}$$

has been liberated.

With a much more refined computation Loper (1978) obtained 2.5×10^{29} joules. If this process of separation has taken 4×10^9 years, the power is

$$\frac{3 \times 10^{29} \text{ joules}}{4.10^9 \times 365 \times 86400 \text{ s}} = \underline{2.4 \times 10^{12} \text{ watts}}$$

which is sufficient to maintain *a non-thermal convection.*

5.4 Some remarks about the boundary layers

The core–mantle interface is a poorly known region inside the Earth. Because of the strong and abrupt discontinuity of all the physical properties (density, rigidity, electrical conductivity, thermal conductivity) at this level, boundary layers are formed and this ensures the transmission of informations from mantle to core and from core to mantle.

Boundary layers appear where a very small parameter (for example the viscosity ν) multiplies the highest differentiated term. This applies too to the advection term and we may introduce the consideration of an *inertial Rossby boundary layer* of thickness proportional to $\varepsilon^{1/2}$.

It again applies for the same reasons to the well-known heat equation

$$\frac{\partial T}{\partial t} + (\bar{q} \cdot \nabla) T = \chi \Delta T \tag{5.8}$$

also written with scaling numbers:

$$\left(\frac{\partial}{\partial t} + \varepsilon \bar{q} \cdot \nabla\right) T = P^{-1} E \Delta T \tag{5.9}$$

so that we have to introduce a *Prandtl thermal boundary layer*.

As a matter of fact the D'' region introduced by Bullen at the base of the lower mantle of the Earth, only about 60 km thick (from 2832 to 2892 km depth) may well be such a thermal boundary layer at the core interface (see Doornbos, 1983). Figure 5.2, taken from Wright *et al.* (1985), shows a net decrease in the speed of P waves across this layer. It could be related to a jump of the thermal conductivity: if χ decreases, dT/dr may increase from 3 K km^{-1} to 13K km^{-1}, that is a variation of 1000 K within 100 km.

The D'' region is supposed by Stacey and Loper (1983) to be the source of material for narrow plumes carrying heat up from the core.

It is thus quite clear that all transport properties involve the existence of boundary layers at any solid–liquid interface.

The relative thickness of such boundary layers is governed by the corresponding dimensionless numbers:

1. the Prandtl number, which is the ratio between the thickness of the viscous boundary layer and the thermal boundary layer;
2. the ratio of Rossby number to Ekman number, equal to the thickness ratio of inertial boundary layer to viscous boundary layer;
3. the Hartmann number or the Prandtl magnetic number should give the ratio between viscous and magnetic boundary layers but we have seen (§4.10) that, in the case of hydromagnetic equations, one has to consider a single Ekman–Hartmann composite boundary layer, the thickness of which depends upon a combination of molecular (or turbulent) viscosity ν and of magnetic viscosity. However we do not know if magnetic viscosity

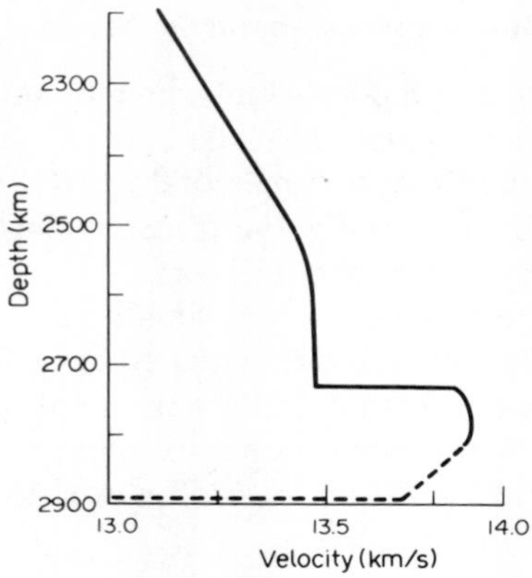

FIGURE 5.2 *P velocity model for the D″* region, at the core-mantle interface, showing a sharp velocity increase of 2.8 % at a depth of 2728 km with a decrease in P wave velocity with increasing depth below 2800 km (according to Wright *et al.*, 1985).

here takes the major role with respect to kinematic (or turbulent) viscosity, because we have no information about the toroidal part of the core magnetic field and no direct information about possible turbulence inside the core. The complications of the magnetic field pattern may cause serious regional disturbances in the thickness of the purely magnetic Hartmann layer.

When a change occurs in the external conditions (for example a variation in the velocity of rotation) the information is transmitted to the internal fluid through the boundary layer mechanism.

The mathematical treatment of the gradients which vary so rapidly inside the boundary layer is made by introducing the stretched variable ξ along the normal to the boundary and by developing $\bar{q}$ and p in powers of $E^{1/2}$. This generates a hierarchy of equations to solve in sequence (see for example Greenspan, 1965 or Busse, 1968).

It is not clear if the Earth's core can be turbulent even if the magnetic charts look like meteorological charts. The turbulence is very difficult to treat, and it may be sufficient here to follow the classical *Boussinesq approach*, which consists in replacing the molecular viscosity v in the Navier equation by a *turbulent viscosity* v^* which may be several orders of magnitude higher than v. It is important to again emphasize here that, while the molecular viscosity is a rheological property of matter, the turbulent viscosity is a property of the movement itself and in no way a property of matter.

In the case of turbulence we will still have a thin laminar boundary layer but in addition a thick turbulent boundary layer where much energy could be

TABLE 5.1 *Observations of Earth Tide diurnal waves*

		Vertical Component			*Horizontal East West Component*		
			$\delta = 1 + h - \frac{3}{2}k$			$\gamma = 1 + k - h$	
Wave	Frequency °/Hour	Amplitude Unit:1μgal	Wahr model	Observed(*)	Amplitude Unit:0″001	Wahr model	Observed(⁑)
$\sigma Q1$							
2Q1	12°9	2.0		1.1569			
σ_1							
Q_1	13°999	6.8	1.152	1.1570	0.57	0.689	0.668 ± 0.075
ρ_1	13°471	1.2		1.1512	0.12	0.689	0.635 ± 0.133
O_1	13°943	35.4	1.152	1.1572	3.31	0.689	0.6735 ± 0.0186
τ_1	14°196	3.0	1.152	1.1596	0.29	0.696	0.700 ± 0.071
NO_1							
π_1	14°918	1.0	1.149	1.1540	0.11	0.699	0.758 ± 0.110
P_1	14°958	16.5	1.147	1.1568	1.67	0.700	0.7098 ± 0.0238
K_1	15°041	49.3	1.132	1.1449	5.29	0.730	0.7497 ± 0.0003
ψ_1	15°082	0.4	1.235	1.2392	0.02	0.523	0.430 ± 0.143
ϕ_1	15°123	0.7	1.167	1.2015	0.08	0.687	0.772 ± 0.117
θ_1	15°513	0.5		1.1759			
J_1	15°585	2.8	1.155	1.1678	0.28	0.692	0.725 ± 0.069
OO_1	16°139	1.5	1.154	1.1631	0.14	0.689	0.662 ± 0.090

h and k are the Love numbers (elastic parameters). Their arithmetic combinations δ and γ are the amplitude factors of the earth tide response.
(*) Ten observing stations
Twelve gravimeters (including 3 Superconducting instruments)
Total of measurements: 398.856 hourly readings.
(⁑) Three underground stations in Belgium and Luxemburg: Dourbes (−45 M), Sclaigneaux (−80 M), Walferdange (−100 M).
Eight Verbaandert-Melchior Quartz Horizontal Pendulums interferometrically calibrated.
Total of measurements: 806.808 hourly readings.
(diurnal waves vanish in North South component at mid latitudes).

dissipated. Would it be possible to observe such phenomena if they exist inside the core?

Most of the dissipation inside the boundary layer occurs around the resonance produced when the container (the mantle) oscillates at a period which is very close to one of the free oscillations of the fluid contained in it (the core).

Thus a very precise determination of the amplitudes of those tidal waves having frequencies sufficiently close to the resonance frequency (i.e. K_1, ψ_1 and ϕ_1 waves) should give very valuable informations (see figure 5.3). The difficulty comes from the very low theoretical amplitude of the wave ψ_1, which is close to resonance, but it may be hoped that superconducting gravimeters will be precise enough to clarify this question before long. Another difficulty is of course related to the oceanic tides contribution, but much progress is under way in this field now.

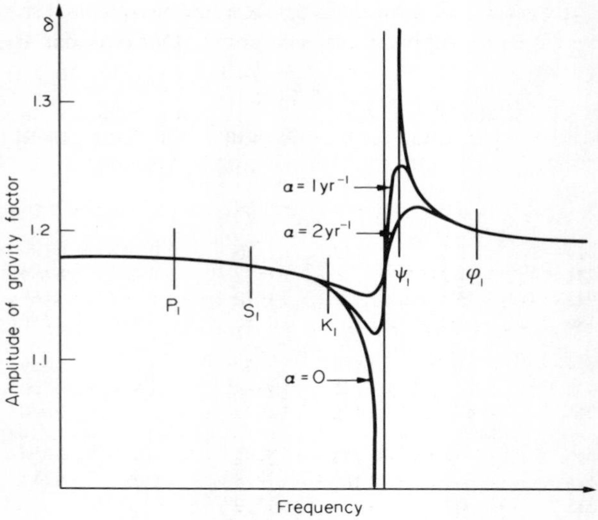

FIGURE 5.3 The viscosity in the liquid Earth core determines the strength of the resonance (according to Sasao, Okamoto and Sakai, 1977).

5.5 Core–mantle coupling

Coupling mechanisms inevitably operate at the core–mantle interface if the core and the mantle do not rotate at the same speed ("spin up") or around the same axis ("spin over"). The problem is to evaluate the efficiency of such couplings.

If the core was perfectly spherical, non-viscous and not magnetic, it could not transfer any movement or information to the mantle and vice-versa: both parts of the Earth should rotate independently.

However, a spherical interface cannot be considered, even as a "first approximation". The first approximation corresponds indeed to the hydrostatic equilibrium, that is to the celebrated Clairaut differential equation (1760) and the most recent integration of this equation gives a core flattening

$$f_c = \frac{1}{392} \tag{5.10}$$

(Denis and Ibrahim, 1981), which corresponds to a difference of 9 km between the polar and equatorial radii of the core.

We can therefore conceive that different coupling mechanisms are simultaneously acting between the core and the mantle:

1. a non-dissipative inertial coupling due to pressure effects related to the geometrical ellipticity of the boundary (9 km) when the axis of rotation of the core differs from the axis of rotation of the mantle;

2. a topographic coupling similarly due to pressure effects if this boundary has an irregular topography (bumps of some 100 m appear as a possibility);
3. a dissipative viscous coupling transmitted through an Ekman boundary layer (10 cm thick or more if turbulent);
4. a dissipative coupling due to magnetic viscosity transmitted through a Ekman–Hartmann boundary layer (meters thick?).

The dissipative couplings tend, of course, to an attenuation of any differential rotation between core and mantle, while the non-dissipative couplings produce permanent gyroscopic effects of the axis of rotation.

These couplings are thought to be responsible for the astronomically observed accelerations of the speed of rotation of the Earth.

Such observed accelerations are of the order of

$$\dot{\Omega} = 1.5 \times 10^{-20}\,\mathrm{rad\,s^{-2}} \tag{5.11}$$

for decade variations and ten times higher for shorter variations.

As the Earth moment of inertia is

$$C = 8 \times 10^{37}\,\mathrm{kg\,m^2} \tag{5.12}$$

this corresponds to torques

$$\begin{aligned} C\dot{\Omega} &= 10^{18}\,\mathrm{Nm} \text{ for decade variations} \\ &\sim 10^{19} \text{ to } 10^{20}\,\mathrm{Nm} \text{ for short period variations} \end{aligned} \tag{5.13}$$

which should be exerted by the envisaged couplings: it is thus essential to evaluate the possible strength of each of these couplings.

It is rather elementary to write the expression of such couplings, by integrating all over the surface of the core-mantle boundary the torque $\overline{R} \wedge \overline{F}$ where $\overline{F}$ represents the force applied at every point of it, either pressure, laminar, or turbulent and magnetic viscous stresses.

In the case of the Lorentz force, however, the coupling is obtained by integrating the torques acting inside the whole volume of the mantle, but Rochester (1962) demonstrated that it can also be reduced to a surface integral (see §5.11).

It is far more difficult to evaluate the strength of each of these couplings in the absence of any precise information about the geometrical characteristics and physical properties of the core–mantle boundary layers.

The inertial coupling

Astronomical observations show that, at the precession frequency $|\overline{\Omega}^*|$ (period 25,800 years), the mantle and the core rotate as if they were rigidly tighted. However the lunisolar gravitational torque which is responsible for precession is proportional to the flattening. It is therefore smaller for the core than for the mantle, the ratio being the ratio of their flattenings: $f_c/f = 298/392 = 0.76$. The pressure torque is therefore required to compensate for this difference. Thus

$$\bar{N}_{in} = \left(1 - \frac{f_c}{f}\right) C_{\text{core}} \bar{\Omega}^* \Lambda \bar{\Omega} \tag{5.14}$$

which, in terms of pressure, can be written:

$$N_{in} = \oiint_S \bar{R} \Lambda (-p\hat{n}) dS = - \iiint_V (\bar{r} \Lambda \text{ grad } p) dV \tag{5.15}$$

This coupling is due to the ellipticity of the boundary ($\bar{R} \Lambda \hat{n} \neq 0$). It is conservative, i.e. without any dissipation, and does not transmit any energy to the dynamo.

If we consider an ellipsoid of revolution as boundary:

$$x^2 + y^2 + \frac{a^2}{c^2} z^2 = a^2 \tag{5.16}$$

and introduce its ellipticity

$$\varepsilon = \frac{a^2 - c^2}{a^2 + c^2}, \quad f = \frac{1 + \varepsilon}{1 - \varepsilon} \tag{5.17}$$

so that

$$x^2 + y^2 + \frac{1 + \varepsilon}{1 - \varepsilon} z^2 = a^2 \tag{5.18}$$

a normal to this surface is

$$\hat{n} = x\hat{x} + y\hat{y} + \frac{1 + \varepsilon}{1 - \varepsilon} z\hat{z} \tag{5.19}$$

while the radius vector is

$$\bar{r} = x\hat{x} + y\hat{y} + z\hat{z} \tag{5.20}$$

Thus

$$\bar{r} \Lambda \hat{n} = 2\varepsilon (y\hat{x} - x\hat{y}) z \tag{5.21}$$

taking

$$\frac{1 + \varepsilon}{1 - \varepsilon} - 1 \sim 2\varepsilon \tag{5.22}$$

The transformation of (5.17) into a Poincaré image sphere is obtained by putting

$$X' = x \qquad Y' = y \qquad Z' = fz. \tag{5.23}$$

Let now

$$\bar{\Omega}_M = \Omega \hat{z} \tag{5.24}$$

be the rotation vector of the mantle and

$$\bar{\Omega}_N = \Omega \hat{z} + \theta \Omega \hat{x} \tag{5.25}$$

be the rotation vector of the core, θ being the angle of separation of both vectors. The components of the velocity in the Poincaré sphere are:

$$-\Omega y \qquad \Omega x - \theta\Omega z \qquad \theta\Omega y \tag{5.26}$$

and, in the ellipsoid, considering eqn (5.23):

$$-\Omega y \qquad \Omega x - \theta\Omega z \qquad f^{-1}\theta\Omega y \tag{5.27}$$

The velocity is thus:

$$\bar{q} = \Omega(x\hat{y} - y\hat{x}) + \theta\Omega\{f^{-1}y\hat{z} - z\hat{y}\} \tag{5.28}$$

from which it follows that

$$\begin{aligned} &\nabla\cdot\bar{q} = 0 \\ \text{and} \\ &\bar{q}\cdot\nabla\bar{q} = -\Omega^2 x\hat{x} - \Omega^2 y\hat{y} + \theta\Omega^2 z\hat{x} + \theta\Omega^2 f^{-1} x\hat{z} \end{aligned} \tag{5.29}$$

by neglecting terms in θ^2.
As $\partial\bar{q}/\partial t = 0$ (ω is constant), one obtains, from the Navier equation:

$$-\frac{1}{\rho}\nabla P = \bar{q}\cdot\nabla\bar{q} \tag{5.30}$$

from which

$$P = \rho\Omega^2\left(\frac{x^2+y^2}{2} - \theta zx\right) \tag{5.31}$$

as $f \approx 1$. Finally, in complex notation:

$$N_{xy} = \frac{2i\varepsilon}{a}\oiint z(x+iy)\,P\,dS \tag{5.32}$$

or

$$N_{xy} = \frac{2i\varepsilon\rho\Omega^2}{a}\oiint z(x+iy)\left(\frac{x^2+y^2}{2} - \theta xz\right)dS \tag{5.33}$$

which, by using spherical coordinates, reduces to

$$N_{xy} = -\frac{8\pi}{15}i\varepsilon\rho\Omega^2\theta R^5 \quad \text{(here } R = a\text{) (Loper, 1975)} \tag{5.34}$$

and

$$\left.\begin{aligned} N_x &= -\frac{8\pi}{15}\varepsilon\rho\Omega^2\theta R^5 \\ N_y &= +\frac{8\pi}{15}\varepsilon\rho\Omega^2\theta R^5 \end{aligned}\right\} \tag{5.35}$$

Still, introducing the kinetic momentum of the core

$$C\Omega = \left(\frac{8\pi}{15}\rho R^5\right)\Omega:$$

$$N_{xy} = i\varepsilon(\theta\Omega)\cdot C\Omega \tag{5.36}$$

which is proportional indeed to the ellipticity ε, to the tilt-over angle θ and to the kinetic momentum. This kind of coupling drastically depends on frequency. It is about zero at the Chandler frequency for which the core does not participate at the movement while, as said before, it is very tight for precession. The liquid core partakes the rigid body motion of the mantle in a narrow range of frequencies.

By comparing eqn (5.14) with eqn (5.34) one can immediately evaluate the spin-over angle θ:

$$\theta = \left(1 - \frac{f_c}{f}\right)\frac{\Omega^* \sin\beta}{\Omega\varepsilon} \tag{5.37}$$

where

$$\Omega^* = 7.74 \times 10^{-12}\,\text{rad s}^{-1}$$
$$\Omega = 7.3 \times 10^{-5}\,\text{rad s}^{-1}$$
$$\sin\beta = \sin 23°27' = 0.398$$
$$1 - (f_c/f) = 0.24$$
$$\varepsilon = 1/400$$

give

$$\theta = 4 \times 10^{-6}\,\text{rad} \sim 0''8 \tag{5.38}$$

and consequently, either from (5.14) or from (5.34):

$$N_{in} = 4.5 \times 10^{20}\,N\cdot\text{m} \tag{5.39}$$

which is sufficient to provide the necessary torque (5.13)

Topographic coupling

This is also a kind of inertial coupling analogous to the torque exerted by zonal winds on the Earth's surface topography.

$$\bar{N}_{\text{top}} = \oiint_S (\bar{r}\Lambda\bar{\bar{\Sigma}})\,dS \tag{5.40}$$

$\bar{\bar{\Sigma}}$ being the tangential shear stress.

Let us imagine, on the core boundary, a rectangular bump of height h with sides x, y ($dS = xy$). The force exerted on it is

$$F = (p_2 - p_1)\,hx \tag{5.41}$$

and the resulting stress

$$\Sigma = \frac{F}{dS} = (p_2 - p_1)\frac{h}{y} \tag{5.42}$$

which should be balanced by the Coriolis force, so that

$$(p_2 - p_1)/y \approx |\hat{t} \cdot 2\rho\bar{\Omega}\Lambda\bar{q}| \tag{5.43}$$

thus

$$\Sigma \approx \rho\Omega qh \qquad \text{(Rochester, 1970)} \tag{5.44}$$

or, taking into account the presence of an Ekman boundary layer:

$$\Sigma = C_T \rho\Omega q(h - \delta) \tag{5.45}$$

C_T is the topographic friction coefficient. With $h \approx 1000$ m, which seems to be an acceptable maximum height for such bumps, $q \sim 10^{-3}$ m s^{-1} and $\rho = 10^4$ kg m^{-3} one has

$$\Sigma \approx 0.7\ \mathrm{N\,m^{-2}}$$

We do not have any *direct* information about bumps at the core–mantle interface but it seems that such a topographic coupling may be efficient.

The viscous coupling

The dissipative coupling can be expressed in terms of the viscous shear stress $\bar{\bar{\Sigma}}$ *at the core*–mantle interface:

$$\bar{N}_v = -\oiint_S \bar{R}\Lambda(\bar{\bar{\Sigma}} \cdot \hat{t})\, dS \tag{5.46}$$

If we write a force of friction in the symbolic form

$$f = C_v \eta^m q^n L^k \tag{5.47}$$

where

$\eta = \rho\nu$ is the dynamical viscosity, dimensions $\{\mathrm{ML^{-1}\,T^{-1}}\}$
q, the velocity, dimensions $\{\mathrm{LT^{-1}}\}$
L, the thickness of the Ekman layer (δ), and
C_v a dimensionless coefficient.

We obtain from

$$\{f\} = \{\mathrm{M\,L\,T^{-2}}\}$$

that

$$m = n = k = 1$$

thus

$$f = C_v \rho\nu\, q\delta \tag{5.48}$$

The viscous stress, which has the dimensions of a pressure, will be f/δ^2, thus

$$\Sigma = C_v \rho v \frac{q}{\delta} \tag{5.49}$$

but

$$\delta = \sqrt{\frac{v}{\Omega}} \quad \text{(eqn 3.235)}$$

$$q = \Omega R$$

thus

$$\Sigma = C_v v^{1/2} \rho \Omega^{3/2} R = C_v \rho \Omega q \delta \tag{5.50}$$

If the tilt angle over $\hat{t}$ is θ and with $dS = R^2 \sin\phi \, d\phi \, d\lambda$, one finally has:

$$N_v \ \therefore \ C_v v^{1/2} \rho \Omega^{3/2} R^4 \theta = C_v E^{1/2} \rho \Omega^2 R^5 \theta \tag{5.51}$$

in terms of the Ekman number.

Stewartson and Roberts (1963) gave a coefficient 4.4* instead of C_v so that with

$$v = 4 \times 10^{-6}\ \text{m}^2\,\text{s}^{-1} \qquad \rho = 10^4\ \text{kg m}^{-3}$$
$$R = 3.5 \times 10^6\ \text{m} \qquad \theta = 4 \times 10^{-6}\ \text{(eqn 5.38)}$$

Loper (1975) obtains

$$N_v = 4.10^{16}\ \text{N m} \tag{5.52}$$

A turbulent viscosity should increase this number by several powers. It may be expected that any topography of the core boundary (bumps) will give rise to turbulence.

According to Toomre (1966) a viscosity of about $10^5\ \text{cm}^2\,\text{s}^{-1}$ (stokes) should be sufficient to couple the core to the mantle in the precession phenomenon independently from the inertial coupling. It should be higher than 10^5 stokes for the 18.6-year nutation and higher than 10^{10} stokes for the monthly nutation.

Electromagnetic couplings

Annual and semiannual changes in the length of the day (l.o.d.) and the excitation of the polar motion are explained to a very large extent by exchanges of angular momentum between the atmosphere and the mantle. One can therefore consider the Earth's core as decoupled from the mantle at periods less than 5 years.

* $\frac{8\pi}{15} \times 2.62 = 4.4.$

On the contrary, longer term variations in the speed of rotation of the Earth appear as an important drift in the difference between the l.o.d. and the atmospheric angular momentum and only the Earth's core is big enough to be considered as the source of such effects by a tight coupling of electromagnetic origin. This necessarily suggests the existence of a non negligible toroidal field, estimated to be ten times the poloidal field i.e. $10^{-2}\,T$ in order that $\bar{j}\Lambda\bar{B}$ will be comparable with $(2\bar{\Omega}\Lambda\bar{q})\rho$.

The correlation shown on Fig. 1.7 is the argument to believe that a large scale fluid circulation in the core—rather than MAC waves—explains the westward drift since no other adequate large source needed to conserve angular momentum is apparently available (Vestine, 1953). As a matter of fact two kinds of electromagnetic coupling are possible, as identified by Roberts (1972) (see also Loper, 1975):

(a) "*leakage coupling*", caused by diffusion into the lower mantle of the electric currents created by the dynamo process in the Earth's core. This results in a Lorentz force applied to the mantle.
(b) "*frozen coupling*" resulting from the hydrodynamic motion $\bar{q}$ inside the core, which redistributes the lines of force emanating from the core surface, inducing currents in the mantle.

Both couples combine with spin-up as well as with spin-over effects so that four kinds of coupling have to be considered.

The *leakage coupling* acting upon the mantle is expressed in terms of the Lorentz force L:

$$\bar{N}_{lm} = \iiint_{\text{Mantle}} (\bar{r}\Lambda\bar{L})\,dV = \iiint_{\text{Mantle}} \bar{r}\Lambda(\bar{j}\Lambda\bar{B})\,dV$$
$$= -\frac{1}{\mu}\oint\!\!\oint_S (\bar{r}\Lambda\bar{B})B_r\,dS \qquad (5.53)$$

Rochester has demonstrated that it can be transformed as follows into the surface integral. In the mantle the velocity q is zero so that the induction equation reduces to

$$\frac{\partial\bar{B}}{\partial t} = -\operatorname{curl}(\eta \operatorname{curl}\bar{B}) \qquad (5.54)$$

while the upper mantle is practically an insulator and shields the lower mantle from the variable external fields.
The magnetic Maxwell stress (4.66)

$$\bar{\bar{\tau}} = \frac{\bar{B}\bar{B}}{\mu} - \frac{|B|^2}{2\mu}\bar{e}_i\bar{e}_i \qquad (5.55)$$

gives

$$\mu \nabla \cdot \bar{\bar{\tau}} = (\bar{B} \cdot \nabla)\bar{B} + \bar{B}(\nabla \cdot \bar{B}) - \nabla \frac{|B|^2}{\mu}$$

$$= \bar{B}(\nabla \cdot \bar{B}) - (\nabla \wedge \bar{B}) \wedge \bar{B} = \bar{j} \wedge \bar{B} \qquad (5.56)$$

(see eqn 4.57 and $\nabla \cdot \bar{B} = 0$).
In the insulating region of the mantle

$$\bar{j} = 0$$

Thus

$$\nabla \wedge \bar{B} = 0 \quad \text{and} \quad \bar{B} = \mu \nabla \psi$$

$$\nabla \cdot \bar{\bar{\tau}} = 0$$

and

$$\iiint_{V_\infty} (\bar{r} \wedge \nabla \cdot \bar{\tau})\, dV = \oiint_{S_m + S_\infty} (\bar{R} \wedge \bar{\tau}) \cdot \hat{n}\, dS = 0 \qquad (5.57)$$

where S_m is the mantle external surface.
At infinity $|\bar{B}|$ tends to zero and this reduces to

$$\iint_{S_m} (\bar{R} \wedge \bar{\tau}) \cdot \hat{n}\, dS = 0 \qquad (5.58)$$

The electromagnetic torque due to the Lorentz force is

$$\bar{N}_{lm} = \iiint_M \bar{r} \wedge (\bar{j} \wedge \bar{B}) = \iiint_M (\bar{r} \wedge \nabla \cdot \bar{\tau})\, dV = \qquad (5.59)$$

$$\oiint_{S_c} \bar{R} \wedge \bar{\tau} \cdot \hat{n}\, dS = \frac{1}{\mu} \oiint_{S_c} \left\{ (\bar{R} \wedge \bar{B})\bar{B} \cdot \hat{n} - \frac{B^2}{2} \bar{R} \wedge \hat{n} \right\} dS$$

Here the core–mantle can be assumed to be spherical ($\bar{R} \wedge \hat{n} = 0$):

$$\bar{N}_{lm} = \frac{1}{\mu} \oiint_{S_c} (\bar{R} \wedge \bar{B})\bar{B}_r\, dS \qquad (5.60)$$

the components of which are:

$$N_1 = \frac{1}{\mu} \oiint_S B_r (B_\lambda \cos\theta \cos\lambda + B_\theta \sin\lambda) R\, dS$$

$$N_2 = \frac{1}{\mu} \oiint_S B_r (B_\lambda \cos\theta \sin\lambda - B_\theta \cos\lambda) R\, dS \qquad (5.61)$$

$$N_3 = -\frac{1}{\mu} \oiint_S B_r B_\lambda R \sin\theta\, dS = -\frac{R^3}{\mu} \int_0^{2\pi} \int_0^{\pi} (B_r B_\lambda)_R \sin^2\theta\, d\theta\, d\lambda$$

$$dS = R^2 \sin\theta\, d\theta\, d\lambda$$

(Rochester, 1962). The last equation can be written

$$N_{lm}^{(u)} = N_3 \approx -\frac{4\pi R^3}{\mu} B_r B_\lambda \tag{5.62}$$

with

$B_r = 3.5$ gauss $= 3.5 \times 10^{-4}$ Weber m^{-2} $= 3.5 \times 10^{-4}$ tesla
$B_\lambda/\xi = 0.35$ gauss $= 0.35 \times 10^{-4}$ Weber m^{-2} (Roberts)
$R = 3.5 \times 10^6$ m
$\mu = 10^{-6}$ henry m^{-1} $= 10^{-6}$ Weber m^{-1} Amp^{-1}

and as

Weber = Volt · s; Volt = Watt Amp^{-1}

Watt · s = joule = N · m

one obtains

$$N_{lm}^{(u)} = 4\pi R^3 \frac{B_r B_\Gamma}{\mu\xi} = 6.6 \times 10^{18}\ \text{N} \cdot \text{m} \tag{5.63}$$

N_3 represents the spin up magnetic leakage torque $N_{lm(u)}$ which may be responsible for the acceleration of the mantle speed of rotation due to jerks like those observed in 1910 and 1969.

The tightness of the coupling is determined by the conductivity in a relatively thin shell of the mantle surrounding the core.

It seems efficient when one considers the correlation between the variations of the Earth's velocity of rotation and the geomagnetic fluctuations, while a time delay of about 10 years is observed (figure 1.7).

To estimate the *frozen coupling* Roberts (1972) considers a fluid motion of velocity $\bar{q}$ at the core surface which creates an electric field $\bar{q}B_r$ and therefore currents of density $j = \sigma q B_r$ in the mantle. The size of the Lorentz force $\bar{j} \Lambda \bar{B}$ is therefore $\sigma q B_r^2$ and the torque can be evaluated as

$$N_{lf}^{(u)} \approx (4\pi R^3) \frac{\sigma q B_r^2}{d} \tag{5.64}$$

where d is the thickness of the conductive layer at the bottom of the lower mantle, that is a skin depth as given by eqn (4.43).
Thus

$$N_{lf}^{(u)} \approx (4\pi R^3) q B_r^2 \tau / \mu d \tag{5.65}$$

with the electrical time constant (Cowling time number)

$$\tau = \mu\sigma d^2 \qquad \eta^* = (\mu\sigma)^{-1} \tag{5.66}$$

If one accepts that $\tau \sim 5$ years and that the velocity q is 0.5 mm s^{-1} similar to the westward drift one obtains again the *desired* strength for this kind of

TABLE 5.2 *The different core–mantle couplings*

Coupling		*Parameters*	*Spin*		*Equation*	*Strength* (*N. m*)	*Authors*
Inertial	N_{in}	(ε)	over (θ)	5.34/5.39	$\left(\frac{8\pi}{15}\rho R^5\right) i\varepsilon\Omega^2\theta$	5×10^{20}	Toomre, 1966 Rochester, 1973
Viscous laminar	N_v	(ν, E)	over (θ)	5.51	$\frac{8\pi}{15}(2.62)\,\nu^{1/2}\rho\Omega^{3/2}R^4\theta$	4×10^{16}	Stewartson–Roberts, 1963
					$= \left(\frac{8\pi}{15}\rho R^5\right)(2.62)\ E^{1/2}\Omega^2\theta$		Loper, 1975
turbulent					estimated	10^{18}?	—
Magnetic leakage	$N_{lm}^{(u)}$	(B_r) (B_λ)	up	5.62/5.63	$(4\pi R^3)\frac{B_r B_\lambda}{\mu}$	$6 \times 10^{18} - 10^{19}$	Rochester, 1970 Stix, 1982
	$N_{lm}^{(0)}$	(B_θ)	over (θ)		estimated	8×10^{16}	Loper, 1975
frozen	$N_{fm}^{(u)}$	(σ)	up	5.65/5.67	$(4\pi R^3)\sigma\frac{q}{d}B_r^2 = (4\pi R^3)qB_r^2\frac{\tau}{\mu d}$	4×10^{18}	Roberts, 1972
	$N_{fm}^{(0)}$	(θ)	over (θ)	5.68	$\left(\frac{\sigma\omega}{\mu}\right)^{1/2} B_r^2 R\theta$	3×10^{16}	Toomre, 1966
					$\left(\frac{8\pi}{15}\rho R^5\right)2^{3/2}(1+i)Ff^{-1/2}B_r^2\Omega^2\theta$		Loper, 1975

Note: $C = \frac{8\pi}{15}\rho R^5$, $\frac{8\pi}{15}(2.62) = 4.4$, F, f are inverse magnetic Reynolds numbers (4.49) or Ekman magnetic numbers. The topographic coupling cannot be estimated.

coupling

$$N_{lf}^{(u)} \sim 4 \times 10^{18}\ \mathrm{N \cdot m} \tag{5.67}$$

This corresponds of course to a "spin-up" effect (the westward drift). The "spin-over" effect has been tentatively estimated by Toomre (1966) to be

$$N_{lf}^{(0)} = \left(\frac{\sigma\omega}{\mu}\right)^{1/2} B_r^2 R^4 \theta \tag{5.68}$$

Note that $\tau = 5$ years $= 1.58 \times 10^8$ s and $\mu\sigma = \eta^{*-1} = 0.5\ \mathrm{m^{-2}\,s}$ give a thickness $d = 18$ km for the skin depth at the bottom of the mantle.

Bibliography

Acheson, D. J. (1975) On hydromagnetic oscillations within the Earth and core–mantle coupling. *Geophys. J. R. Astr. Soc.*, **43**, 253–68.

Alterman, Z., Jarosch, H. and Pekeris, C. L. (1959) Oscillations of the Earth. *Proc. Roy. Soc.*, **A252**, 80–95

Anufriyev, A. P. and Braginskiy, S. I. (1975) Influence of irregularities of the boundary of the Earth's core on the velocity of the liquid and on the magnetic field. *Geomag. Aeronomy*, **15**, 754–7.

Anufriev, A. P. and Braginsky, S. I. (1983) The influence of irregularities of the core–mantle boundary on Earth core hydrodynamics. *Magnetic Field and the Processes in the Earth's Interior*. Czechoslovak Acad. Sciences, 368–73.

Bolt, Bruce A. and Niazi, Mansour (1984) S velocities in D″ from diffracted SH-waves at the core boundary. *Geophys. J. R. astr. Soc.* **79**, 825–834.

Bondi, H. and Gold, T. (1950) On the generation of magnetism by fluid motion. *Monthly Notices RAS*, **110**, 607–11.

Braginsky, S. I. (1971) Origin of the geomagnetic field and its secular change, *IAGA Bull.*, **31**, 41–54.

Braginskiy, S. I. and Fishman, V. M. (1976) Electromagnetic interaction of the core and mantle when electrical conductivity is concentrated near the core boundary. *Geomag. Aeronomy*, **16**, 443–46.

Bukowinski, M. S. T. and Hauser, J. (1980) Pressure-induced metallization of SRO and BAO theoretical estimate of transition pressures. *Geophys. Res. Lett.*, **7**(9), 689–692.

Busse, F. H. (1968) Shear flow instabilities in rotating systems. *J. Fluid Mech.*, **33**(3), 577–89.

Crossley, D. J. and Rochester, M. G. (1980) Simple core undertones. *Geophys. J. R. Astr. Soc.*, **60**, 129–61.

Davies, P. A. (1978) Topographic effects in rotating stratified fluids: laboratory experiments. *Rotating Fluids in Geophysics*. Academic Press, 249–91.

Denis, C. and Ibrahim, A. (1981) On a self-consistent representation of Earth models, with an application to the computing of internal flattening. *Bull. Geod.*, **55**, 179–95.

Doornbos, D. J. (1978) On seismic-wave scattering by a rough core–mantle boundary. *Geophys. J. R. astr. Soc.* **53**, 643–662.

Doornbos, D. J. (1983) Present seismic evidence for a boundary layer at the base of the mantle. *J. Geophys. Res.*, **88**, 3498–505.

Doornbos, D. J. and Mondt, J. C. (1979) P and S waves diffracted around the core and the velocity structure at the base of the mantle. *Geophys. J. R. astr. Soc.* **57**, 381–395.

Dungey, J. W. (1950) A note on magnetic fields in conducting materials *Proc. Camb. Phil. Soc.*, **46**, 651–4.

Elsasser, W. M. (1949) Non-uniformity of the Earth's rotation and geomagnetism, *Nature*, **163**(4140), 351–2.

Eltayeb, L. A. and and Hassan, M. H. A. (1979) On the effects of a bumpy core–mantle interface. *Phys. Earth Plan. Int.*, **19**, 239–54.

Fearn, D. R. and Loper, D. E. (1981) Compositional convection and stratification of Earth's core. *Nature*, **289**, 393–94.

Feber, R. C., Wallace, T. C. and Libby, L. M. (1984) Uranium in the Earth's core. *EOS*, **65**(44), 785.

Friedlander, S. (1985) Internal oscillations in the Earth's fluid core. *Geophys. J. R. Astr. Soc.*, **80**, 345–61.

Garland, G. D. (1957) The figure of the Earth's core and the non-dipole field. *J. Geophys. Res.*, **62**, 486–7.

Greenspan, H. P. (1965) On the general theory of contained rotating fluid motions. *J. Fluid Mech.*, **22**(3), 449–62.

Gubbins, D., Thomson, C. J. and Whaler, K. A. (1982) Stable regions in the Earth's liquid core. *Geophys. J. R. Astr. Soc.*, **68**, 241–51.

Hage, H. and Muller, G. (1979) Changes in dimensions, stresses and gravitational energy of the Earth due to crystallisation at the inner-core boundary under isochemical conditions. *Geophys. J. R. Astr. Soc.*, **58**, 495–508.

Hassan, M. H. A. and Eltayeb, I. A. (1982) On the topographic coupling at the core–mantle interface. *Phys. Earth Plan. Int.*, **28**, 14–26.

Hide, R. (1966) Free hydromagnetic oscillations of the Earth's core and the theory of the geomagnetic secular variation *Phil. Trans. Roy. Soc. Lond.* **A259**, 615–47.

Hide, R. (1969) Interaction between the earth's liquid core and solid mantle. *Nature*, **222**, 1055–1056.

Hide, R. (1970) On the Earth's core–mantle interface. *Q. J. Roy. Meteor. Soc.*, **96**(410), 579–87.

Hide, R. (1977) Towards a theory of irregular variations in the length of the day and core–mantle coupling. *Phil. Trans. Roy. Soc. Lond.* **A284**, 547–54.

Hide, R. (1982) On the role of rotation in the generation of magnetic fields by fluid motions. *Phil. Trans. R. Soc. Lond.*, **306**, 223–34.

Hide, R. and Horai, K. (1968) On the topography of the core–mantle interface. *Phys. Earth Plan. Int.*, **1**, 305–8.

Hide, R. and Malin, S. R. C. (1970) Novel correlations between global features of the Earth's gravitational and magnetic fields. *Nature*, **225**, 605–9.

Hide, R. and Malin, S. R. C. (1981) On the determination of the size of the Earth's core from observations of the geomagnetic secular variation. *Proc. Roy. Soc. Lond.* **A374**, 15–33.

Houben, H. and Turcotte, D. L. (1975) Can the geomagnetic dynamo be driven by the semi-diurnal tides? *EOS*, **56**, 356.

Jacobs, J. A. (1975) *The Earth's Core*. Academic Press.

Jeanloz, R. and Richter, F. M. (1979) Convection, Composition, and the Thermal State of the Lower Mantle. *J. Geophys. Res.* **84**, 5497–5504.

Jones, G. M. (1977) Thermal Interaction of the Core and the Mantle and Long-Term Behavior of the Geomagnetic Field. *J. Geophys. Res.* **82**, 1703–1709.

Le Mouel, J. L. and Courtillot, V. (1981) Core motions, electromagnetic core–mantle coupling and variations in the Earth's rotation—new constraints from geomagnetic secular variation impulses. *Phys. Earth Plan. Int.*, **24**, 236–41.

Longuet-Higgins, M. S. (1966) Planetary waves on a hemisphere bounded by meridians of longitude. *Phil. Trans. Roy. Soc. Lond.*, **260**, 318–50.

Loper, D. E. (1975) Torque balance and energy budget for the precessionally driven dynamo. *Phys. Earth Plan. Int.*, **11**, 43–60.

Loper, D. E. (1978) The gravitationally powered dynamo. *Geophys. J. R. Astr. Soc.*, **54**, 389–404.

Loper, D. E. (1984) The dynamical structures of D″ and deep plumes in a non-Newtonian mantle. *Phys. Earth Plan. Int.* **34**, 57–67.

Loper, D. E. and Roberts, P. H. (1981) A study of conditions at the inner core boundary of the Earth. *Phys. Earth Plan. Int.*, **24**, 302–7.

Loper, D. E. and Roberts, P. H. (1983) Compositional convection and the gravitationally powered dynamo. *Stellar and Planetary Magnetism* (ed. Soward) in *Fluid Mechanics of Astrophysics and Geophysics*, vol. 2, 297–327. Gordon and Breach.

Malin, S. R. C. and Hide, R. (1982) Bumps on the core–mantle boundary. Geomagnetic and gravitational evidence revisited. *Phil. Trans. Roy. Soc. Lond.*, **306**, 281–9.

Moffatt, H. K. and Dillon, R. F. (1976) The correlation between gravitational and geomagnetic fields caused by interaction of the core–mantle interface. *Phys. Earth Plan. Int.*, **13**, 67–78.

Roberts, P. H. (1972) Electromagnetic core–mantle coupling. *J. Geomag. Geoelectr.*, **24**, 231–59.

Roberts, P. H. (1974) Magnetic core–mantle coupling by inertial waves. *Phys. Earth Plan. Int.*, **8**, 389–90.

Rochester, M. G. (1960) Geomagnetic westward drift and irregularities in the Earth's rotation. *Phil. Trans.*, **A252**, 531–54.

Rochester, M. G. (1962) Geomagnetic core–mantle coupling. *J. Geophys. Res.*, **67**(12), 4833–6.

Rochester, M. G. (1970) Core–mantle interactions: geophysical and astronomical consequences. In *Earthquake displacement fields and the rotation of the Earth.* Astrophysics and Space Science Library, ed. L. Mansinha, D. E. Smylie and A. E. Beck, D. Reidel, Dordrecht.

Rochester, M. G., Jacobs, J. A., Smylie, D. E. and Chong, K. F. (1975) Can precession power the geomagnetic dynamo. *Geophys. J. R. Astr. Soc.*, **43**, 661–678.

Roden, R. B. (1963) Electromagnetic core–mantle coupling. *Geophys. J. R. Astr. Soc.*, **7**, 361–74.

Ruff, Larry, J. and Helmberger, Don, V. (1982) The structure of the lowermost mantle determined by short-period P-wave amplitudes. *Geophys. J. R. astr. Soc.* **68**, 95–119.

Sasao, T., Okamoto, I. and Sakai, S. (1977) Dissipative core–mantle coupling and nutational motion of the Earth. *Publ. Astr. Soc. Japan*, **29**, 83–105.

Shankland, T. J. and Brown, J. M. (1985) Homogeneity and temperatures in the lower mantle. *Phys. Earth Plan. Int.* **38**, 51–58.

Smith, M. L. (1976) Translational inner core oscillations of a rotating, slightly elliptical Earth. *J. Geophys. Res.*, **81**, 3055–65.

Stacey, F. D. and Loper, D. E. (1983) The thermal boundary layer interpretation of D″ and its role as a plume source. *Phys. Earth Plan. Int.*, **33**, 45–55.

Stewartson, K. and Roberts, P. H. (1963) On the motion of a liquid in a spheroidal cavity of a precessing rigid body. I. *J. Fluid Mech.*, **17**, 1–20.

Stix, M. (1982) On electromagnetic core–mantle coupling. *Geophys. Astr. Fluid Dyn.*, **21**, 303–13.

Stix, M. and Roberts, P. H. (1984) Time-dependent electromagnetic core–mantle coupling. *Phys. Earth Plan. Int.*, **36**, 49–60.

Szeto, A. M. K. and Smylie, D. E. (1984) Coupled motions of the inner core and possible geomagnetic implications. *Phys. Earth Plan. Int.*, **36**, 27–42.

Toomre, A. (1966) On the coupling of the Earth's core and mantle during the 26000-year precession. *The Earth–Moon System.* Plenum Press, 33–45.

Tresl, J. (1977) Magnetohydrodynamic oscillations due to precessional motion. *Studia Geophys. Geod.*, **1**, Series 21, 27–34.

Tresl, J. (1983) Precession of the Earth and the internal geomagnetic field. *Magnetic Field and the Processes in the Earth's Interior.* Czechoslovak Acad. Sciences, 382–8.

Verhoogen, J. (1981) Heat balance of the Earth's core. *Geophys. J. R. Astr. Soc.*, **4**, 276–81.

Verhoogen, J. (1973) Thermal regime of the Earth's core. *Phys. Earth Plan. Int.*, **7**, 47–58.

Veronis, G. (1970) The analogy between rotating and stratified fluids. *Annual Rev. Fluid Mech.* **2**, 37–67.

Vestine, E. H. (1953) On variations of the geomagnetic field, fluid motions, and the rate of the earth's rotation. *J. Geophys. Res.* **58**, 127–45.

Watanabe, H. and Yukutake, T. (1975) Electromagnetic core–mantle coupling associated with changes in the geomagnetic dipole field. *J. Geomag. Geoelectr.*, **26**, 153–73.

Wright, C., Muirhead, K. J. and Dixon, A. E. (1985) The P wave velocity structure near the base of the mantle. *J. Geophys. Res.*, **90**(81), 623–34.

Appendix

Correspondence of units currently used in internal geophysics

	SI system	*CGS system and others*
Force	Newton: $N = m\,kg\,s^{-2}$	dyne $= 10^{-5}$ N
Pressure	Pascal: $Pa = Nm^{-2}$	bar $= 10^{6}$ dyne cm^{-2} $= 10^{5}$ Pa
Acceleration	$m\,s^{-2}$	gal $= 10^{-2}\,m\,s^{-2}$; 0.1 μgal $= 1\,nm\,s^{-2}$
Energy, work, moment	Joule: $J = N \cdot m$	erg $= 10^{-7}$ joule calorie = 4.1868 joule
Power	Watt: $W = J\,s^{-1}$	
Kinematic viscosity	$m^2\,s^{-1}$	Stoke $= cm^2\,s^{-1}$ $= 10^{-4}\,m^2\,s^{-1}$
Dynamic viscosity	Pascal-second: Pa s	poise $= 10^{-1}$ Pa s
Resistivity	ohm: $\Omega = V\,A^{-1}$	
Conductivity	Siemens = ohm^{-1}: $S = \Omega^{-1}$:	1 siemens $= 10^{-11}$ emu
Magnetic induction flux	Weber: $Wb = V \cdot s$	
Magnetic induction	Tesla: $T = Wb\,m^{-2}$ $= kg\,s^{-2}\,A^{-1}$	gauss $= 10^{-4}$ T gamma $= 10^{-9}$ T
Inductance	Henry: $H = Wb\,A^{-1}$ $= \Omega S$	
Thermal conductivity	Watt $metre^{-1}$ $Kelvin^{-1}$	
Heat flow	Watt m^{-2}	heat flow unit: h f u = $41.8 \times 10^{-3}\,W\,m^{-2}$ $= 10^{-6}\,cal\,cm^2\,s^{-1}$
Activation Enthalpy	Joule $mole^{-1}$ (1 mole $= 6.02 \times 10^{23}$ atoms, Avogadro number	100 kilocalories per mole $= 4.1868 \times 10^{5}$ joules per mole $= 7 \times 10^{-12}$ erg per atom = 4 eV)

See: Payne, M. A. (1981). SI and Gaussian CGS units, conversion and equations for the use in geomagnetism *Phys. Earth Plan. Int.*, **26**, 10–16.

1. The Earth

Speed of light	299,792,458 m s^{-1}
Equatorial radius a	6,378,137 ± 1.5 m
Polar radius c	6,356,752 m
Surface	5.1×10^{14} m^2
Equivolume sphere radius $\sqrt{a^2c}$	6,371,000 m
Geometrical flattening $f = (a-c)/a$	$1/298.257222101 = 3.335281 \times 10^{-3}$
Ellipticity $(a^2-c^2)/(a^2+c^2)$	1/297.75
Volume $(4\pi/3)a^2c$	1.083207×10^{21} m^3
Dynamical form factor $J_2 = \dfrac{C-A}{Ma^2}$	$10826.3 \pm 0.1 \times 10^{-7}$
$\dot{J}_2$	-3×10^{-11} year^{-1}
Precession constant $H = \dfrac{C-A}{C}$	$1/305.43738 = 0.003273993$
C/Ma^2 ratio	0.330676
GM	$3,986,005 \pm 0.5 \times 10^8$ m^3 s^{-2}
G	6.6726 ± 0.0005 $\times 10^{-11}$ kg^{-1} m^3 s^{-2}
Mass M	5.97369×10^{24} kg
Mean density	5.5148×10^3 kg m^{-3}
Moment of inertia $C = 0.330676\ Ma^2$	8.036559×10^{37} kg m^2
Moment of inertia A	8.010247×10^{37} kg m^2
Angular velocity Ω	7.292115×10^{-5} rad s^{-1}
Angular momentum $C\Omega$	5.86035×10^{33} kg m^2 s^{-1}
Normal gravity at the equator	9.7803267 m s^{-2}
Normal gravity at the poles	9.832186 m s^{-2}
$q = \Omega^2 a^3/GM$	$3,461392196 \times 10^{-3}$
Precession angular velocity	7.74×10^{-12} rad s^{-1}
Moment of inertia of the Mantle C	7.185×10^{37} kg m^2

2. The Earth's core

Geometrical and mechanical characteristics

External boundary radius	$(3485 \pm 2) \times 10^3$ m = 0.547 of the Earth
Internal boundary radius	$(1217 \pm 10) \times 10^3$ m
Hydrostatic flattening at external boundary	1/392.15 or $a - c = 9 \times 10^3$ m
Surface of external boundary	1.526×10^{14} m^2
Surface of internal boundary	0.186×10^{13} m^2
Volume	1.697×10^{20} m^3 = 0.157 of the Earth
Mean density	1.1×10^4 kg m^{-3}
Mass	1.867×10^{24} kg = 0.312 of the Earth
Moment of inertia C	8.508×10^{36} kg m^2
Diurnal rotation velocity Ω	7.292115×10^{-5} rad s^{-1}
Angular momentum $C\Omega$	6.2041×10^{32} kg m^2 s^{-1}
Coriolis force $2\Omega a$ (at the pole)	5.0826×10^2 m s^{-1}
Total volume (with inner core)	1.773×10^{20} m^3

	At external boundary	*At internal boundary*	
Density	0.99	1.22	$\times 10^4$ kg m^{-3}
Density of liquid pure iron	—	1.29	$\times 10^4$ kg m^{-3}
Gravity	1068	440	gal = cm s^{-2}
Pressure	1.37	3.27	$\times 10^{11}$ Pa
Incompressibility	7	10	$\times 10^{11}$ Pa
Rigidity	$< 10^6$		Pa
Viscosity	10^{-6}		m^2 s^{-1}
Q	$\sim 10^6$		

(10^{11} Pa = 1 megabar = 10^{12} dyne cm^{-2})
1 stokes = 10^{-4} m^2 s^{-1}

3. Thermodynamic properties of the Earth's core

Temperature	3200–4200 K
Adiabatic gradient of temperature	~ 0.5 K km^{-1}
Adiabatic heat flow	4.10^{12} watts
Grüneisen parameter	~ 1.4
Specific heat C_p	500 to 700 J kg^{-1} K^{-1} = 5×10^6 erg g^{-1} K^{-1}
Latent heat of fusion	4.10^5 J kg^{-1} = 4×10^9 erg g^{-1}

Thermal conductivity k	$\sim$ 39–60 watts $m^{-1} K^{-1}$
Thermal expansion α	$\sim 2 \times 10^{-5} K^{-1}$
Thermal diffusivity κ	$\sim 4.2 \times 10^{-2}$ stokes $= 4.2 \times 10^{-6} m^2 s^{-1}$
Kinematic viscosity ν	$\sim 10^{-2}$ stokes $= 10^{-6} m^2 s^{-1}$

4. Electromagnetic properties of the Earth's core

Electrical resistivity $\eta^* = (\mu\sigma)^{-1}$	2×10^{-6} ohm·m
Electrical conductivity σ	5 to 7×10^5 siemens m^{-1}
Magnetic diffusivity λ	$3 \times 10^4 cm^2 s^{-1} = 3 m^2 s^{-1}$
Poloidal Magnetic dipole B_p	$\sim$ 4 gauss
Toroidal field B_T	? 100 to 200 gauss?
Dipole westward drift	$0.05°$ year$^{-1} \sim 1.8 \times 10^{-4} m s^{-1}$
Non dipole westward drift	$0.2°$ year$^{-1} \sim 7 \times 10^{-4} m s^{-1}$
Flow speed	$0.03 cm s^{-1}$
Alfven speed	$0.4 cm s^{-1}$ to $10 cm s^{-1}$ (depending whether $B = 4$ to 100 gauss)
Magnetic energy $\frac{1}{2\mu}\int_V \lvert B\rvert^2 dV$	$\sim 6 \times 10^{22}$ joules

5. Dimensionless numbers in the Earth's core

Ekman number	$2 \times 10^{-10} \rightarrow 10^{-15}$
Rossby number	$4 \times 10^{-6} \rightarrow 4.10^{-7}$
Reynolds number	2×10^7
Nusselt number	7×10^2
Prandtl number	0.27–1.00
Péclet number	3×10^7
Rossby magnetic number	2×10^{-9}
Reynolds magnetic number	$10^2 \rightarrow 10^4$
Prandtl magnetic number	$1.5 \times 10^{-6} \rightarrow 3 \times 10^{-8}$
Hartmann number	4.6×10^6
Froude number	6.6×10^{-4}

6. The Earth's inner core

Radius a	1217×10^3 m $= 0.191$ of the Earth
Volume	$7.5502 \times 10^{18} m^3 = 0.007$ of the Earth
Surface	$0.186 \times 10^{13} m^2$
Mean density	$12 \times 10^3 kg m^{-3}$
Mass M	9.0603×10^{22} kg $= 0.0015$ of the Earth
$C = \frac{2}{5}Ma^2$	$5.3676 \times 10^{33} kg m^2$

Angular momentum $C\Omega$	3.914×10^{29} kg m^2 s^{-1}
Gravity at the surface	4.4 m s^{-2} = 440 gal
Temperature	~ 4000 K
Specific heat C_p	~ 660 J kg^{-1} K^{-1}
Thermal conductivity k	~ 35 W m^{-1} K^{-1}
Thermal expansion α	~ 8×10^{-6} K^{-1}

Differential operators and fundamental formulas

$\bar{q} = u\hat{\imath} + v\hat{\jmath} + w\hat{k}$

$\nabla V = \text{grad } V$

$\nabla \cdot \bar{q} = \text{div } \bar{q}$

$$\nabla \wedge \bar{q} = \text{curl } \bar{q} = \left(\frac{\partial w}{\partial y} - \frac{\partial v}{\partial z}\right)\hat{\imath} + \left(\frac{\partial u}{\partial z} - \frac{\partial w}{\partial x}\right)\hat{\jmath} + \left(\frac{\partial v}{\partial x} - \frac{\partial u}{\partial y}\right)\hat{k}$$

$\nabla^2 V = \text{lap } V = \Delta V$

div grad V = lap V	$\nabla \cdot \nabla V = \Delta V$
curl grad $V = 0$	$\nabla \wedge \nabla V = 0$
div curl $\bar{q} = 0$	$\nabla \cdot (\nabla \wedge \bar{q}) = 0$
lap V = grad div V − curl curl V	$\Delta V = \nabla(\nabla \cdot V) - \nabla \wedge (\nabla \wedge V)$

$\frac{1}{2} \text{grad } (\bar{v} \cdot \bar{v}) = \bar{v} \wedge \text{curl } \bar{v} + \bar{v} \cdot \text{grad } \bar{v}$

$\frac{1}{2} \nabla v^2 = \bar{v} \wedge (\nabla \wedge \bar{v}) + (\bar{v} \nabla) \bar{v}$

$\frac{1}{2} \text{grad } (\bar{\Omega} \wedge \bar{r})^2 = \bar{\Omega} \wedge (\bar{\Omega} \wedge \bar{r})$

$\text{curl } (B \cdot \bar{a}) = B \text{ curl } \bar{a} - \bar{a} \wedge \text{grad } B$

$\text{div } (\rho \bar{V}) = \rho \text{ div } \bar{V} + \bar{V} \text{ grad } \rho$

$\text{curl } (\bar{A} \wedge \bar{B}) = \bar{B} \text{ grad } \bar{A} - \bar{A} \text{ grad } \bar{B} + \bar{A} \text{ div } \bar{B} - \bar{B} \text{ div } \bar{A}$

$\text{div } (\bar{A} \wedge \bar{B}) = \bar{B} \text{ curl } \bar{A} - \bar{A} \text{ curl } \bar{B}$

$\text{curl } \phi \bar{A} = \phi \text{ curl } \bar{A} + (\text{grad } \phi) \wedge \bar{A}$

Solenoidal field: $\nabla \cdot \bar{q} = 0$

Any vector field may be divided into a curl free and a divergence free (solenoidal) parts:

$$\bar{Q} = -\text{grad } \psi + \text{curl } \bar{A}$$

Stokes theorem

The work of vector $\bar{A}$ along a closed contour C (circulation of $\bar{A}$) is equal to the flux of its curl across any surface S sustained by C:

$$\oint_C \bar{A} \cdot d\bar{x} = \oiint_S (\text{curl } \bar{A}) \cdot \hat{n} \, ds$$

Green–Ostrogradsky theorem

$$\oiint_S \bar{B} \cdot \hat{n} \, ds = \iiint_V \mathit{div} \, \bar{B} \, dv$$

Vectorial operations

$\vec{a}\cdot(\vec{b}\wedge\vec{c}) = \vec{b}\cdot(\vec{c}\wedge\vec{a}) = \vec{c}\cdot(\vec{a}\wedge\vec{b}) = (\vec{a}\wedge\vec{b})\cdot\vec{c}$

$\vec{a}\wedge(\vec{b}\wedge\vec{c}) = (\vec{a}\cdot\vec{c})\,\vec{b} - (\vec{a}\cdot\vec{b})\,\vec{c}$

$(\vec{a}\wedge\vec{b})\cdot(\vec{c}\wedge\vec{d}) = (\vec{a}\cdot\vec{c})\,(\vec{b}\cdot\vec{d}) - (\vec{a}\cdot\vec{d})\,(\vec{b}\cdot\vec{c})$

$(\vec{a}\wedge\vec{b})\wedge(\vec{c}\wedge\vec{d}) = \vec{b}\,(\vec{a}\cdot(\vec{c}\wedge\vec{d})) - \vec{a}\,(\vec{b}\cdot(\vec{c}\wedge\vec{d}))$

Coriolis force in spherical coordinates

Let us put, in spherical coordinates

$$\bar{q} = u\hat{r} + v\hat{\theta} + w\hat{\lambda}$$

with

$$\bar{\Omega} = (\Omega\cos\theta)\hat{r} - (\Omega\sin\theta)\hat{\theta}$$

and

$$\hat{r}\wedge\hat{\theta} = \hat{\lambda} \quad \hat{\theta}\wedge\hat{\lambda} = \hat{r} \quad \hat{\lambda}\wedge\hat{r} = \hat{\theta}$$

One obtains

$$2\bar{\Omega}\wedge\bar{q} = -(2w\Omega\sin\theta)\hat{r} - (2w\Omega\cos\theta)\hat{\theta} + (2v\Omega\cos\theta + 2u\Omega\sin\theta)\hat{\lambda}$$

Legendre polynomials

$$P_n^m(\mu) = \frac{(1-\mu^2)^{m/2}}{2^n\, n!}\left(\frac{d}{d\mu}\right)^{m+n}(\mu^2-1)^n$$

with $\mu = \cos\theta$, θ being the colatitude.

Theorem of reciprocity (Kelvin)

Considering the radius

$$r = (x^2+y^2+z^2)^{1/2}$$

one has

$$\nabla^2 r^m = m(m+1)r^{m-2}$$

thus

$$\nabla^2 r^{-1} = 0$$

On the other hand,

$$\Delta(UV) = U\Delta V + V\Delta U + 2\left(\frac{\partial U}{\partial x}\frac{\partial V}{\partial x} + \frac{\partial U}{\partial y}\frac{\partial V}{\partial y} + \frac{\partial U}{\partial z}\frac{\partial V}{\partial z}\right)$$

and, if V_n is an harmonic function of degree n ($V_n = r^n S_n(\theta, \lambda)$; S_n being a surface harmonic function) and $U = r^m$, this gives

$$\Delta (r^m V_n) = m(2n + m + 1) r^{m-2} V_n$$

and, consequently

$$\Delta (r^{-2n-1} V_n) = 0$$

Thus, if $V_n = r^n S_n\ (\theta, \lambda)$ is an harmonic function of degree n, then $r^{-2n-1} V_n = r^{-n-1} S_n\ (\theta, \lambda)$ is an harmonic function of degree $-n-1$.

Differential equations of the second order

Elliptic equation

Example: the Laplace equation

$$\frac{\partial^2 W}{\partial x^2}+\frac{\partial^2 W}{\partial y^2}+\frac{\partial^2 W}{\partial z^2}=0$$

This can be solved if values of the function are known on the boundary. There are no wave form solutions.

Parabolic equation

Example: the heat equation, diffusion equations

$$\frac{\partial^2 W}{\partial x^2}-\frac{1}{\alpha}\frac{\partial W}{\partial t}=0$$

This can be solved if an initial value and a value on the boundary is known. There are no wave form solutions. Heat transfer gives confused sensorial perception.

Hyperbolic equation

Example: the vibrating string equation

$$\frac{\partial^2 W}{\partial x^2}-\frac{1}{c^2}\frac{\partial^2 W}{\partial t^2}=0$$

This can be solved if the displacement and velocity are known at $t = 0$. There are wave form solutions. Typical is sound propagation; it gives very acute sensorial perceptions.

The physics of planetary interiors involves the consideration of these three forms of the partial second-order differential equations: the elliptic Laplace equation in the potential theory, the parabolic heat equation in thermodynamics and the hyperbolic wave equation in all problems of oscillations.

Plane waves

In a bidimensional case the incompressibility allows to introduce a stream function such as

$$u=\frac{\partial\psi}{\partial y}\qquad v=-\frac{\partial\psi}{\partial x}$$

so that one obtains the typical equation:

$$\frac{\partial}{\partial t}(\nabla^2\psi)+\beta\frac{\partial\psi}{\partial x}=0$$

for which we can try a solution

$$\psi = \exp i(lx + my - \sigma t)$$

This represents a plane wave which has a constant phase

$$\theta = lx + my$$

all along the straight lines of equation:

$$lx + my = \text{constant}$$

which we may call crest lines.

$$\vec{k} = \text{grad}\,\theta = l\hat{i} + m\hat{j}$$

is called the *wave number* and

$$k = |\bar{k}| = \sqrt{l^2 + m^2}$$

is the number of times one crosses the value 2π along a *unit* distance orthogonal to the crest lines. Then

$$\lambda = \frac{2\pi}{|\bar{k}|}$$

is the wave length and

$$l = k \cos \alpha \qquad m = k \sin \alpha$$

α being the angle between $\vec{O}x$ and $\vec{k}$.

The *frequency* is the number of crest crossings per unit of time:

$$-\frac{\partial \theta}{\partial t}$$

Now, putting the solution into the equation gives

$$\sigma(l^2 + m^2) + \beta l = 0$$

that is

$$\frac{\sigma}{l} = -\frac{\beta}{l^2 + m^2} = -\frac{\beta}{k^2}$$

$\frac{\sigma}{l}$ represents the velocity of a crest along the Ox axis.

The phase velocity is

$$c = -\frac{1}{\nabla\theta}\frac{\partial \theta}{\partial t} = \frac{\sigma}{|\bar{k}|} = \frac{\sigma}{\sqrt{l^2 + m^2}}$$

but from

$$\sigma = -\beta l/(l^2 + m^2)$$

we obtain the components of the group velocity

$$\frac{\partial \sigma}{\partial l} = \frac{\beta(l^2 - m^2)}{(l^2 + m^2)^2} = \frac{\beta}{k^2} \cos 2\alpha$$

$$\frac{\partial \sigma}{\partial m} = \frac{\beta 2lm}{(l^2 + m^2)^2} = \frac{\beta}{k^2} \sin 2\alpha$$

so that the group velocity vector makes an angle 2α with the x axis

$$\bar{c}_g = \left(\frac{\partial \sigma}{\partial l}, \frac{\partial \sigma}{\partial m} \right)$$

Toroidal and poloidal vectorial fields

Let $\hat{r}$, $\hat{\theta}$, $\hat{\lambda}$ be a local referential upon a spherical surface. The vectorial field $\bar{s}(r, \theta, \lambda)$ can be decomposed as follows (Love):

$$\bar{s}(r,\theta,\lambda) = \hat{r}U(r,\theta,\lambda) + \text{curl}\{\bar{r}\,W(r,\theta,\lambda)\} + \bar{r}\Lambda\,\text{curl}\{\bar{r}\,V(r,\theta,\lambda)\}$$

Because of the spatial geometry of the problem one introduces spherical harmonics for the scalar functions (U, V, W). As

$$\text{curl}(\bar{r}W) = W\,\text{curl}\,\bar{r} - \bar{r}\Lambda\,\overline{\text{grad}}\,W = -\bar{r}\Lambda\,\overline{\text{grad}}\,W$$

$$\Delta W = \text{grad div}\,W - \text{curl curl}\,W = 0$$

and

$$r = \begin{vmatrix} r\hat{r} \\ 0 \\ 0 \end{vmatrix}, \quad \text{grad}\,W = \begin{vmatrix} \dfrac{\partial W}{\partial r}\hat{r} \\ \dfrac{\partial W}{r\partial\theta}\hat{\theta} \\ \dfrac{\partial W}{r\sin\theta\partial\lambda}\hat{\lambda} \end{vmatrix}, \quad \begin{array}{l} \hat{r}\Lambda\hat{\theta} = +\hat{\lambda} \\ \hat{r}\Lambda\hat{\lambda} = -\hat{\theta} \end{array}$$

one has

$$\text{curl}(\bar{r}W) = -\hat{r}\Lambda\,\overline{\text{grad}}\,W = \hat{\theta}\frac{\partial W}{\sin\theta\partial\lambda} - \hat{\lambda}\frac{\partial W}{\partial\theta}$$

and

$$\bar{r}\Lambda\,\text{curl}(\bar{r}V) = \hat{\lambda}\frac{\partial V}{\sin\theta\partial\lambda} + \hat{\theta}\frac{\partial V}{\partial\theta}$$

Let us define the gradient operator on the sphere as

$$\nabla_S = \hat{\theta}\frac{\partial}{\partial\theta} + \hat{\lambda}\frac{\partial}{\sin\theta\partial\lambda}$$

one can then write the vectorial field $\overline{S}$ either as

$$\bar{s}(r,\theta,\lambda) = \hat{r}U + \hat{\theta}\left(\frac{\partial V}{\partial\theta} + \frac{\partial W}{\sin\theta\partial\lambda}\right) + \hat{\lambda}\left(\frac{\partial V}{\sin\theta\partial\lambda} - \frac{\partial W}{\partial\theta}\right)$$

either as

$$\bar{s} = \hat{r}U + \nabla_S V - \hat{r}\Lambda\nabla_S W$$

$\hat{r}U$	represents a radial displacement;
$\hat{r}U + \nabla_S V$	represents the spheroidal deformation;
$-\hat{r}\Lambda\nabla_S W$	represents the torsional deformation.

In the analysis of free oscillations one uses the more explicit notation:

spheroidal deformations:

$$ {}_n\bar{S}_l^m = \hat{r}\{{}_nU_l(r)\,Y_l^m(\theta,\lambda)\} + \bar{\nabla}_S\{{}_nV_l(r)\,Y_l^m(\theta,\lambda)\} $$

toroidal deformations:

$$ {}_n\bar{T}_l^m = {}_nW_l(r)\{-\hat{r}\Lambda\bar{\nabla}_S Y_l^m(\theta,\lambda)\} $$

where n is the number of roots in the U, V, W functions of r which define the "overtones". It is clear that:

$$ \bar{S}_l^m = \hat{r}\Lambda\,\bar{T}_l^m \qquad \bar{T}_l^m = -\hat{r}\Lambda\bar{S}_l^m $$

Glossary

Activation energy
Energy needed to allow an imperfection in a crystal to jump the potential barrier which separates it from another minimum energy position.

Advection
Effect of non-linear term in the Navier–Stokes equation $(\bar{q}\cdot\nabla)\bar{q}$.

Alfvén waves
Shear waves propagating along the magnetic lines of force as in the case of an elastic harp string.

Baroclinic mode
Surfaces of constant pressure are inclined over surfaces of constant density (internal waves result). Hydrostatic equilibrium cannot generally be maintained.
Calculating the strength of a vortex tube as

$$\frac{d\bar{\Gamma}}{dt} = \oint_c \frac{d\bar{q}}{dt}\cdot d\bar{r} = -\oint_c (2\bar{\Omega}\Lambda\bar{q})\cdot d\bar{r} - \oint_c \frac{\nabla p}{\rho}\cdot d\bar{r}$$

from the Navier equation, the last term can be written:

$$-\oint_c \frac{\nabla p}{\rho}\cdot d\bar{r} = -\iint_A \operatorname{curl}\left(\frac{\nabla p}{\rho}\right)\cdot\bar{n}\,dA = \iint_A \frac{\nabla\rho\Lambda\nabla p}{\rho^2}\cdot\bar{n}\,dA$$

$\dfrac{\nabla\rho\Lambda\nabla p}{\rho^2}$ is the baroclinic vector.

Barotropic mode
Surfaces of constant pressure coincide with surface of constant density (surface waves can occur).

Boussinesq approximation
The changes of density resulting from the motions are supposed to be due to thermal effects rather than to pressure and, because the thermal expansion coefficient is very small, these changes can be neglected in all terms except the buoyancy term.

Buoyancy
Results from variations of density inside a fluid subject to gravity: it is due to the Archimedean force.

Conservative forces
No dissipation, no hysteresis.

Cowling constant
Characteristic time related to the skin depth d in geomagnetism ($= \mu\sigma d^2$).

Dispersion
When the wave velocity depends on its frequency.

Dynamical form factor
The ratio $(C - A)/Ma^2$ where C and A are the polar and equatorial moments of inertia.

Ergodic hypothesis
For a stationary process, the temporal mean and the spatial mean are identical.

Eutectic
Chemical mixture melting or solidifying at a temperature which is lower than the melting temperatures of its constituants.

Frozen magnetic field
When there is no dissipation the magnetic lines of force move with the fluid as if it was frozen.

Geostrophic equilibrium
The Coriolis force is balanced by the pressure gradient, the other terms in the Navier equation being negligible.

Isentropic
The changes of state are so rapid that transport processes can be ignored.

Isothermal
The heat conduction is so efficient that the temperature of a fluid particle remains constant.

Magnetic jerk
Acceleration or impulse in the secular variation of the Earth's magnetic field. It happened in the late 1960s but also around 1840 and 1905.

Magnetostrophic equilibrium
The Coriolis force is balanced by the Lorentz force.

Maxwell pressure and shear stress
The Lorentz force can be decomposed into two terms corresponding to an isotropic pressure and a shear stress.

Neutral equilibrium
The weight of a fluid element is exactly balanced, at every point, by the pressure exerted on it by neighbouring fluid elements.

Skin depth
The distance at which the amplitude of a periodic oscillation is reduced by a factor $1/e$.

SNREI model
Spherical, Non-Rotating, Elastic, Isotropic model.

Solenoidal field
Divergence equal to zero.

Taylor columns
By applying the curl operator to the geostrophic Navier equation one obtains motions in columns parallel to the axis of rotation.

Turbulence
Loss of memory of the previous situation.

Author Index

Subject Index

1. Famous equations and laws

2. Dimensionless numbers

Hydrodynamics

Magnetism

3. Miscellaneous subjects

Waves